THE ICE AGE
WORLD

THE ICE AGE WORLD

An Introduction to Quaternary
History and Research with Emphasis
on North America and Northern Europe
During the Last 2.5 Million Years

By

Bjørn G. Andersen, University of Oslo
Harold W. Borns Jr., University of Maine

SCANDINAVIAN UNIVERSITY PRESS
Oslo – Copenhagen – Stockholm

Scandinavian University Press (Universitetsforlaget AS),
PO Box 2959 Tøyen, N-0608 Oslo, Norway

Distributed world-wide excluding Scandinavia by
Oxford University Press, Walton Street, Oxford OX2 6DP

Oxford   New York   Toronto
Dehli   Bombay   Calcutta   Madras   Karachi
Kuala Lumpur   Singapore   Hong Kong   Tokyo
Nairobi   Dar es Salaam   Cape Town
Melbourne   Auckland   Madrid
and associated companies in Berlin   Ibadan

Oxford is a trade mark of Oxford University Press

Published in the United States
by Oxford University Press Inc., New York

© Scandinavian University Press (Universitetsforlaget AS) 1994

ISBN 82-00-21810-4

Published with grants from:
    The Research Council of Norway
    Norsk Hydro A.S
    Saga Petroleum AS
    Statoil

British Library Cataloguing in Publication Data
Data available

Library of Congress Cataloguing in Publication Data
Data available

Design: Hans Nusslé
Printed in Norway by Emil Moestue as, Oslo 1994

Acknowledgements

We are most grateful for the support from the
following Norwegian institutes and companies:

THE RESEARCH COUNCIL OF NORWAY

NORSK HYDRO A.S

SAGA PETROLEUM A.S.

STATOIL

Without their generous financial contributions it
would not have been possible to publish this book.

Bjørn G. Andersen   *Harold W. Borns Jr.*

Extended Acknowledgements

We are very grateful to numerous people who have rendered information and in various ways helped us with the manuscript. Professor David Smith checked Chapters 1, 2 and 3. The following professors/scientists have corrected illustrations or shorter sections of text: Jan Mangerud, John Birks, Kaare Høeg, Per Aagaard, Odd Nilsen, Svein Manum, Torgeir Andersen, Kari Henningsmoen, Ingrid Olsson, Burchard Menke, Georg Jacobsen, Herbert Wright Jr., and Calvin Heusser. We are also grateful to all colleagues who generously put their illustrations/photographs at our disposal. In particular, we wish to thank the Department of Geology at the University of Oslo for good support, Grete Andersen for patiently typing the manuscript, and the drawing office and photo laboratory of the University of Oslo for assistance with several illustrations.

The manuscript was submitted in the spring of 1992.

Table of Contents

INTRODUCTION

Rapid and dramatic changes have taken place during the last 2.5 million years of the earth's history. Changes in climate have caused large ice sheets to expand and contract. Land areas increased far into present sea areas as the global sea level periodically dropped and rose 100 m to 150 m in response to the growth and melting of glaciers. Continents and islands that today are separated by ocean water were then connected by land bridges. Large parts of present-day desert areas were covered with vegetation where herds of animals lived, or they were covered with large lakes. A strange Arctic fauna with wooly elephant (mammoth), wooly

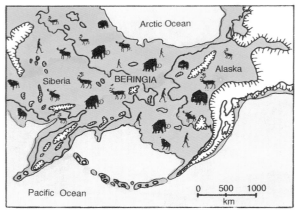

Fig. 1–1. Matterhorn in Switzerland, the "Queen of Mountains", was sculptured by ice age glaciers. The Matterhorn is a small remnant of a much larger mountain which was surrounded by glaciers that attacked on all sides. Even the existing small glaciers attack and erode the mountain slopes, together with expansion of cracks in the rocks due to the freezing and thawing of water, which breaks up and disintegrates the hard rock surface. The large volume of rock debris on the slopes and glaciers on both sides is being transported rapidly downslope, at the same time as new rock debris is formed on the mountain slopes. In this way Matterhorn is gradually changing, and it is doomed to be removed by these natural processes within a short time, geologically speaking.

The general alpine landscape shown on the picture is typical of many glacially sculptured mountain landscapes around the world, such as in the Rocky Mountains, the Andes, the Himalayas, and the Southern Alps of New Zealand. They are eroded by alpine glaciers. (Photo by Beat Perren, Air Zermatt.)

Fig. 1–2. The Bering Land Bridge during the last ice ages. The Bering Land Bridge area, Beringia, was an extensive emerged land area connecting Asia and America. Animals and plants migrated across this land bridge. Early humans also migrated eastwards, but probably as late as 14 000–12 000 years ago when the bridge was somewhat narrower than indicated. The map shows the approximate maximum extent of Beringia during the last glaciation, and the presented Alaskan glaciers represent their maximum extent at the height of the last glaciation, the Wisconsin/Weichsel Glaciation. Note that the glaciers at this stage (about 18 000 years ago) blocked the passage from Beringia to the southern parts of North America. (Modified from D. Hopkins, 1967.)

Fig. 1–3. Europe, 21 000–18 000 years ago. The North European Ice Sheet reached as far south as Berlin and Warsaw, and still further south during earlier ice ages. Tundra (orange) and steppe/parkland (yellow) covered most of Europe. An exotic ice age fauna with mammoth, wooly rhinoceros, and reindeer lived on the tundra and steppe/parkland all the way south to the Mediterranean, where early humans also lived and hunted. Present-day Europe received much of its surficial sediment cover during this ice age. The glacier left a blanket of glacial deposits, and strong winds, partly generated over the glacier, spread sand and fine-grained silt/loess over much of the tundra and steppe beyond the glacier. The climate was extremely cold, particularly during winters and in areas adjacent to the ice sheet and over the pack ice that covered part of the North Atlantic. The illustration shows winter-ice conditions on the ocean. For the present North Sea area, the geographic reconstruction records conditions 21 000–22 000 years ago (see p. 55). Pictures of mammoth and wooly rhinoceros are painted by Zenêk Burian. Photos by Reinhard Tierfoto (muskox) and J. Östeng How (reindeer).

Fig. 1–4. The farmlands of central Jutland, Denmark, lie on sediments transported by the ice age glaciers and glacial meltwater rivers. Much of the farmlands in northern Europe and North America lie on sediments of similar origin. (Photo by E.W. Olsson – Luft Foto.)

The hummocky terrain, with scattered small lakes to the right of the dashed line, represents a section of the Mid-Jutland end-moraine complex which was deposited along the ice front about 20 000–18 000 years ago. The flat, partly forested terrain to the left of the dashed line is underlain by a sandy and gravelly outwash plain, deposited by meltwater rivers from the ice sheet. The cover of glacially and glaciofluvially deposited sediments is up to several tens of meters thick.

rhinoceros, reindeer, muskox, and bison lived on the tundra and grassland which covered most of Europe and parts of North America south of the large ice sheets. And, finally, the last 2.5 million years saw the evolution of modern humans.

The importance of the geological deposits from this period cannot be overemphasized. They are the soft sediments upon which rich soils develop and vegetation grows, and they cover much of the land areas in significant parts of the world. Over large areas these deposits are of glacial or periglacial origin (see Fig. 1–4). This is true for much of the excellent farm land and the wooded areas in North America, the former Soviet Union, Europe, Argentina, Chile, New Zealand, and parts of Asia. Several of the large glaciers that existed in, for instance, North America and in Europe had a tremendous erosive force, as well as a power to transport and deposit eroded rock material. Spectacular evidence of this glacial erosion can be seen, for example, in the alpine areas of the Rocky Mountains, the Alps, and the Himalayas, as well as the fjords of Norway, Alaska, and Chile.

Several of the world's large oceans were partly covered with sea ice, ice shelves, and occasionally even with grounded ice sheets during the cold glacial periods of the last 2.5

million years. Sediment layers studied in long cores taken from the ocean floors reveal an almost unbelievably continuous and detailed record of global climate and glacier fluctuations.

For several decades the authors have been teaching introductory courses about the history of the ice age world, and have felt the need of a book with a brief, general introduction to this subject focussing on the geological and geographical aspects, which represent the basic themes of Quaternary research. For practical purposes the content of this book is divided in four parts: *1. historical review; 2. outline of the ice age history, mainly in North America and northern Europe; 3. introduction to important processes and the scientific methods used to document this history; and 4. an expanded glossary of terms and concepts used in this book, along with other useful and important information for undergraduate students of Quaternary geology.*

The main text in the book presents an adequate background for an introductory course at the university level. For more interested readers we have added lists of books and presented comments about the books which we think (a) provide good and easily read reviews of specialized subjects, and (b) provide a good scientific treatment of the subject, but which require a more advanced scientific

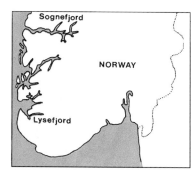

Fig. 1–5. Lysefjord in southwest Norway.
The fjords are examples of enormous erosion and sculpturing by the ice age glaciers, and Lysefjord is a classic fjord occupying a rock trough, which is about 457 km deep in the central parts and only 50 m deep near its mouth. However, a large glacially deposited gravel ridge, an end moraine, crosses the fjord on the shallow rock threshold at the mouth, and the top of the end moraine is only 10–15 m deep. The Lysefjord is carved into hard crystalline bedrock, granites, and gneisses. Only glaciers can erode rock basins of this kind. Observe that the picture shows only the inner half of the fjord, while the bathymetric map and the longitudinal profile show the entire fjord. The plateaus on both sides lie about 1000 m above sea level. The green-colored ridges are young end moraines, about 10 500 and 9500 years old.
Note also that Lysefjord, although typical in form, is very small compared to the largest fjords in western Norway. For instance, the Sognefjord has a rock floor about 1500 m deep with a 300 m deep rock threshold near its mouth. (Published with permission from Lyse Kraft. The bathymetric map is modified from Kåre Strøm, 1936.)

background. Many terms in the text are explained in the Glossary, and students should familiarize themselves with this and use it when encountering new words.

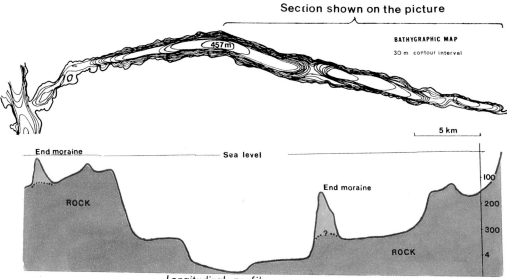

Chapter 1
HISTORICAL REVIEW

Fig. 1–6. Erratic in Jotunheimen Mountains, central Norway. The erratic is of a dark-colored gabbro and it rests on a mountain top of light-colored quartzite. According to the legend the erratics were thrown by giants who lived in Jotunheimen ("the home of giants"). In this case the erratic must have been thrown several km, from the closest bedrock outcrop of the gabbro. Today we know that the erratic was transported by a glacier.

Evolution of the glacial theory

The theory about the presence and history of former large ice sheets, the so-called glacial theory, is rather new, and the development of how the theory emerged is exciting. Several geological features which we now recognize as typical of glacial origin had been observed for a long time before the glacial theory evolved, and their origin was explained in many unique and interesting ways. For example, the origin of huge boulders found resting on unweathered smooth bedrock surfaces on mountain tops and elsewhere stirred the imagination. They were frequently of a different rock type from that of the bedrock upon which they rested, so they were obviously transported to their present location, and they were so large that no human beings could have transported them. But how did they get there?

Evidence of active giants

In the 18th century some countries, like Norway, were thought by some people to have been heavily populated with giants (trolls) which lived in the mountains. They were blamed for throwing rocks at each other when they fought, and apparently the trolls were not always friendly, as evidenced by the numerous large boulders on the mountain tops (Fig. 1–6).

Fig. 1–7A. Close-up of Trollgaren.

Fig. 1–7B. Trollgaren, "the giant fence", on the mountain plateau near Jösenfjord in southwest Norway, about 800 m above sea level. According to the legend Trollgaren was built by the Jösenfjord Giant (Troll) to keep the neighboring giant out of his domain. Today we know that Trollgaren is a moraine ridge deposited at the margin of a glacier about 9500 years ago. The glacier lay to the left of the wall. (Photo by Norsk Fly og Flyfoto.)

They also built huge stone walls to keep neighboring trolls out of their domains. Trollgaren ("the giant wall") found crossing the 800 m high mountain plateau in southwestern Norway is this kind of wall (Fig. 1–7). Today we know that this wall is a part of a marginal moraine deposited along the margin of a glacier about 9500 years ago.

Flood theories

Most scientists did not believe in giants, but many did believe in the biblical deluge. During the 18th century and far into the 19th century respectable scientists explained the transported boulders, which were called erratics, as rocks that had been transported by a flood that passed over even the highest mountains. However, many scientists found that transport of large boulders by water up to the high mountain tops was at best problematic. In 1830

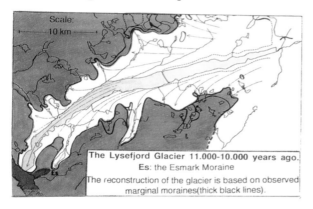

the Scottish geologist Charles Lyell "solved" this problem by suggesting that the boulders had been deposited by large icebergs that floated in the water of the flood. Sediments, including the large boulders, which he visualized had drifted with the flood, were called "drift", a term still applied to these deposits, which are now attributed to glacial action. This growing body of knowledge led to the replacement of the diluvial theory by the drift theory.

Fig. 1–8. The Esmark Moraine in southwest Norway is a 20–30 m high end-moraine ridge deposited 11 000–10 000 years ago at the front of a branch from the Lysefjord Glacier. The ridge, which is locally called Vassryggen, spans the valley floor and dams Lake Haukelivann. A wide outwash plain deposited by meltwater rivers from the glacier lies in front of the moraine. The steep fjord side of Lysefjord is seen in the background. Professor J. Esmark described this ridge in 1824, and by comparison with similar ridges and deposits at the fronts of existing glaciers in Jotunheimen in central Norway, he concluded that this is a glacially deposited end moraine. He was the first to postulate a glaciation for northern Europe. The location is shown on Fig. 2–6. (Photo on top by Norsk Fly og Flyfoto.)

Fig. 1–9. Large erratics on the Saalian moraine west of Berlin in central Germany are practically all of Fennoscandian origin. Erratics from Aaland (A) and Dalane (D) were identified by the writer. It was this kind of erratic which A. Bernhardi observed when he, in 1832, concluded that a polar ice cap reached as far as southern Germany.

Fig. 1–10. The Findeln Glacier and Valley near Zermatt, Switzerland. The features created by this and many other glaciers in Switzerland are so striking that it is no wonder that it was in this country the glacial theory was born. Figures A, B, C, D and E are close-ups of some of the most striking glacial features at Findeln Glacier. The same kind of features are also observed at many sites far from the existing glaciers in Switzerland, in areas which were glaciated during the ice ages. M-M is a sharp lateral-moraine ridge which was deposited during the last few centuries, the upper part as late as 1850, when the Findeln Glacier filled the valley up to the crest of the moraine ridge.

A: The lateral-moraine ridge, M-M. The present-day glacier is depositing a new lateral moraine at lower levels to the right.

B: A part of the ice front showing large boulders on the ice surface. Some of them are as much as 3 m in diameter (see person below x for scale). Several boulders (erratics) are clearly glacially striated; see the large one in front of the glacier. The erratics were therefore transported near the base of the glacier before they were sheared up to the glacier surface, near the glacier front. In a relatively short time the erratics on the ice surface will "slide" down to the foot of the glacier where numerous erratics dumped in previous years are resting. Horizontal arrows mark distinctive shear planes.

C: A section through the top of the lateral moraine showing a typical unsorted glacial deposit, a till, with large erratics of which the largest is about 2 m in diameter.

D: Glacially striated bedrock formed underneath a larger Findeln Glacier. The striated rock surface passes underneath the existing glacier. The rock surface on the entire valley floor in front of the glacier is spectacularly striated. x: the ice-cored end moraine; y: the glacier surface.

E: A small end-moraine ridge deposited about 1875, either at a stationary glacier front or at the front of an advancing glacier which pushed the ridge up.

The glacial theory, in which drift is explained as deposits of former large glaciers, was developed in the Alps during the latter part of the 18th and the early part of the 19th century. In the Alps glacial features can easily be observed adjacent to existing glaciers, and striking glacial features, including large transported granite boulders resting upon glacially striated and polished limestone bedrock surfaces, can be observed in areas far beyond the existing glaciers, as, for example, in the Jura Mountains. The obvious conclusion that the boulders were transported by former, more expansive glaciers was reached by both laymen and scientists such as the Swiss lawyer B.F. Kuhn in 1787, the Scottish scientists J. Hutton in 1795 and J. Playfair in 1802, and the Swiss mountaineer J.P. Perraudin in 1815. Between 1816 and 1833 the Swiss engineer I. Venetz presented substantial evidence of greatly expanded glaciers in the Alps, and in 1829 he suggested that thick glaciers had extended across the Jura Mountains and northwards onto the European plain. This view was presented in a lecture at the Society of the Hospice of the Great St. Bernard, and in 1834 the Swiss scientist J. de Charpentier supported this theory in a lecture to the Swiss Society of Natural Sciences. However, even earlier, in 1824, the Norwegian geologist J. Esmark had presented convincing evidence for an expanded glaciation in northern Europe (Fig. 1–8), and in 1832 the German scientist A. Bernhardi described cobbles and boulders of Scandinavian origin found in Germany, and suggested that a "polar ice cap" reached as far south as southern Germany (Fig. 1–9). In 1837 the botanist K. Schimper, who studied erratics in Bavaria, introduced the term "Die Eiszeit" in a poem to commemorate Galilei's birthday. He also presented notes to the Swiss scientist Louis Agassiz, in which he suggested that most of Europe, Asia and North America had been covered with thick ice. Agassiz, a specialist on fossil fish, attended Charpentier's lecture in 1834, and he objected strongly to his conclusions, but in 1836 he joined Charpentier in the field and became converted to the glacial theory. Thereafter he was an enthusiastic defender of this theory, and he shocked the scientific community, first in 1837 in a presentation to the Swiss Society of Natural Sciences and later (1840) in the famous publication "Etudes sur les glaciaires", where he postulated that ice

A **B**

C

Fig. 1–11. Glacially striated bedrock surfaces are characteristic of formerly glaciated regions. Notice the striking similarity with the striations at Findeln Glacier, Fig. 1–10.

A: Rock surface striated by the West-Antarctic Ice Sheet in the Ellsworth Mountains, at a time when the surface of the ice sheet was at least 400–500 m higher than at present.

B: Glacially striated rock surface in southernmost Norway, about 180 km from the closest existing glacier.

C: Glacial striation near Gothenburg, Sweden, more than 300 km from the closest existing glacier.

The bedrock has been shaped into "whale-backs" by the ice flow. They have gentle stoss-slopes and abrupt, steep lee-slopes. The whale-back shape also is characteristic for glacially sculptured rocks. This shape is particularly well developed on the rock in picture B, where the glacier moved towards the left.

Fig. 1–12. Erratic of a grey-colored granite which rests on a pink-colored granite in Acadia National Park, on the coast of Maine, USA. The grey granite outcrops in an area 30–40 km north of Acadia, and it must have been transported this distance by the glacier. The closest present-day glacier lies more than 1000 km from Acadia. (Photo by Peter L. Kresan.)

A

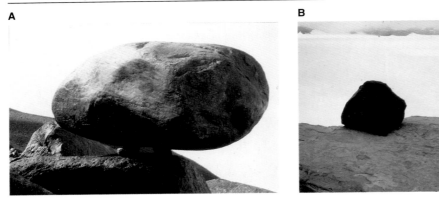

B

C

Fig. 1–13. Erratic boulders (transported by glaciers) lie everywhere near existing glaciers. The pictures show erratics which lie on top of nunataks, 400–500 m above the surfaces of Antarctic glaciers, such as Beardmore Glacier (A+C) and the West-Antarctic Ice Sheet at Ellsworth Mountains (B). Note that the erratic on picture A is perched, and that it rests on three small erratics. The large erratics are from 2–3 m in diameter.

sheets of a distant age stretched from the North Pole to the shores of the Mediterranean. Even believers in the glacial theory thought that Agassiz went too far in his vision, and indeed he did. However, he became so active in presenting evidence of glaciations both in Europe and North America, and advocated a version of the ice age theory so vigorously,

that he frequently has been called a glacial evangelist, or the father of the glacial theory.

The disagreement between believers in the glacial theory and believers in Lyell's flood theory was hard fought throughout most of the 19th century. However, the evidence in favor of the glacial theory and against the flood theory was so strong that towards the end of the century most serious scientists adopted the glacial theory to explain what we now know are the products of glacial erosion and deposition.

The present is the key to the past

In the late 18th century James Hutton and Charles Lyell conceived the thesis that *the present is the key to the past*, a thesis which became a guide in all fields of natural science, including glacial geology. An intensive study of present-day processes like the behavior of glaciers began in the 19th century, and it has accelerated during the last few decades. Features that we know are unique to glacial activity can be observed within all regions of former glaciations, and conversely they are missing in areas which have not been glaciated. In fact, the regional distribution of the glacial features is used to outline the extension of the former glaciers (see Figs. 1–10 to 1–13).

Raised marine shorelines and isostasy

Marine shorelines and marine sediments lying well above present sea level are striking features found along the coasts of many of the formerly glaciated regions, for example, Fennoscandia, Scotland, and North America. High-lying marine features in Scandinavia, and particularly those in the Baltic Sea region, were described during the 17th and 18th centuries by many observers, including the famous Swedish scientists A. Celsius (1724) and C. von Linné (1734). Most of them suggested that the high-lying features resulted from a drop in world sea level, and some related them to a drop following the biblical flood. In 1765–69 the Finnish scientist E. Runeberg postulated that the features could have resulted from a depression and a following rise of the earth's crust, and in 1802 the Scottish scientist

Fig. 1–14. A raised marine terrace near Alta, northern Norway. The large terrace lies about 70 m above sea level, and it is composed of outwash sand and gravel deposited in front of a glacier which occupied the valley behind the terrace 11 000–10 500 years ago. Corresponding shorelines in this area are very distinctive, and it was these that A. Bravais described in 1838. He reported that they were tilted, and in fact, they drop more than 1 m per km seawards and lie only 10 m above sea level on the outermost coast (see Figs. 1–15A and 3–49B). The location of Alta is shown on Fig. 1–15A.

J. Playfair suggested that the earth's surface could have been locally uplifted as a result of heat expansion in deeper parts of the earth below the uplifted areas. A similar theory was adopted by the German geologist L. von Buch (1806–8). However, the theory which generally involved the biblical flood and subsequent drop in sea level was still the one widely accepted, and the defense of the theory engaged both scientists and laymen, including the famous German author Goethe, who in 1822 attacked L. von Buch for his defense of the uplift theory.

In 1838 the French scientist A. Bravais observed that two high-lying shorelines in northern Norway are actually tilted (Fig. 1–14). However, the first scientists to relate the raised marine shoreline features to glaciation were probably the French mathematician J. Adhemar (1842) and the Scottish scientist C. Maclaren (1841–42). Adhemar suggested that ocean water was gravitationally pulled towards the heavier ice-loaded parts of the globe, and Maclaren related high-lying shorelines to changes in sea level caused by the fluctuations in the volume of the ice age glaciers, the so-called glacio-eustatic changes. However, the most important step was probably made in 1865 when the Scottish scientist T. Jamieson attributed the shore features in formerly glaciated regions to the unloading of the heavy ice sheets and the corresponding isostatic rise of the earth's crust. Intensive studies of the raised shorelines showed that they are tilted and rise towards centers which correspond with the centers of the former large ice sheets. In North America this center lies in the Hudson Bay region; in northern Europe it lies in the northern Baltic Sea region. In both regions the highest shorelines are approximately 300 m above present sea level (Figs. 1–15, 1–16).

In an attempt to explain how the ice load caused the depression of the crust, Jamieson suggested that the earth's crust rests on a layer which is in a state of "fusion" and will yield under pressure. This was the introduction of the modern isostasy theory, which explains the "floating" balance between the rigid upper part of the earth, the lithosphere, and the underlying viscous astenosphere (see p. 165). Loading or unloading on the earth's surface results in a corresponding vertical sinking and rising of the lithosphere. This down and up

A

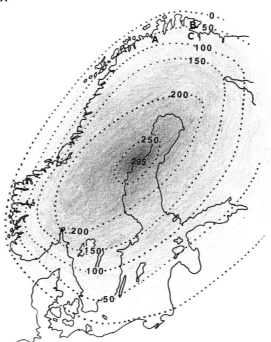

Fig. 1–15A. The elevation, in meters, of the highest raised marine features (terraces, shorelines, beach ridges) in different parts of Fennoscandia, presented by isolines. The highest marine features of the entire region lie about 295 m above sea level in the northern Baltic Sea area. The center and thickest part of the Fennoscandian Ice Sheet also occurred in this area, which was therefore the area of maximum isostatic depression and hence the area of greatest subsequent isostatic uplift. A: Alta. B + C: Location of sites shown in Figs. 1–15C and 1–22A. (Modified from E. Granlund and G. Lundqvist, 1949.)

B

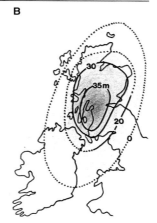

Fig. 1–15B. The isostatic uplift in Great Britain was relatively small since the last "big" ice sheet which covered the area, about 20 000–17 000 years ago, was fairly thin. The center of uplift in Scotland corresponds with the center of the ice sheet.

Fig. 1–15C. Beach ridges at Varangerfjord, northern Norway (see Fig. 1–15A). The uppermost ridge, R-R, is about 13 000–13 500 years old, and it represents the "marine limit" (ML), which is here about 55 m above sea level. (Photo by Fjellanger-Wideröe A/S.)

C

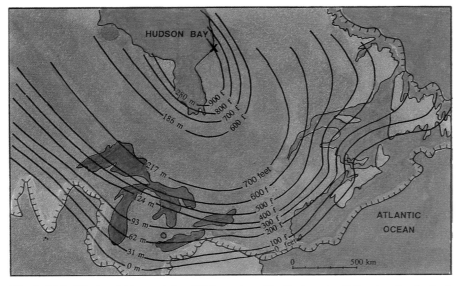

Fig. 1–16A. The isostatic crustal uplift in eastern North America following the deglaciation of the Laurentide Ice Sheet. The contour lines (isolines) connect points with equal uplift during the ice-free (postglacial) period. They are based on the observed highest shore features. The features in the Hudson Bay area are younger than the features in the Great Lakes region. The presented uplift represents the apparent uplift, and the eustatic rise of sea level must be added to obtain the true isostatic uplift. The hachured green line represents the approximate maximum extent of the Laurentide Ice Sheet during the Late Wisconsin. Blue: The Atlantic Ocean and Hudson Bay. Dark blue: The Great Lakes. x: location of Fig. 1–16B. (Modified from R. F. Flint, 1971.)

Fig. 1–16B. Isostatically raised beach ridges adjacent to Richmond Gulf in Hudson Bay, Canada (see Fig. 1–16A). The highest ridges in this area are about 8000 years old, and they lie nearly 300 m above sea level. (Photo by C. Hillaire-Marcel.)

motion of the rigid lithosphere to reach a floating balance is the essence of the isostasy theory.

The concept of multiple glaciations

Another concept which surprised the scientific community was the postulation of more than one glacial period. In 1822 I. Venetz recorded an organic lignite layer between two beds with

glacial deposits near the shore of Lake Geneva in Switzerland, and he suggested that the beds represented two glacial periods separated by a warm period. In the mid-19th century several scientists recorded two separate drift beds, which they correlated with two glaciations: E. Collumb in the Vosges Mountains, J. Trimmer in East Anglia, A.C. Ramsay in Wales, R. Chambers in Scotland, and A. Morlot in Switzerland. Slightly later, in 1863, A. Geikie presented evidence of a warm period separating two glacial advances in Scotland, based upon a warm-climate fossil flora in sediments sandwiched between two glacial till beds.

Towards the end of the 19th and the beginning of the 20th century, many scientists presented evidence of several Quaternary glaciations separated by warm interglacials: four glaciations in the British Isles, three glaciations in most of northern Europe, four glaciations in North America, and at least four glaciations in the Alps were recognized (Fig. 2–1). In particular Penck and Brückner's studies (1901–9) in the Alps were important, and the concept of the four major Quaternary glaciations and three major warm interglacials which was established at this time dominated scientific thinking throughout much of the early half of the 20th century, although several scientists continued to be sceptical of many of the correlations upon which the theory of four-fold glaciation was based.

One serious problem which geologists who study sedimentary beds in stratigraphic sections on land always face is the problem of discontinuous records. Periods of sedimentation were frequently followed by periods of erosion when pages of the historical record were removed. Another problem is concerned with dating the events represented by the observed layers. Throughout the first half of the 20th century the dating methods used were in general primitive, and the correlation of beds from one area to another was problematical. This led, in many cases, to correlations which were not scientifically well founded. Still, in spite of all these problems, the fact remained that there had clearly been several cold glacial periods separated by deglacial warm phases. This concept was accepted by most serious scientists dealing with Quaternary stratigraphy and history. However, today we know that there were several more than four glaciations.

New scientific techniques opened a new era of Quaternary research

During the last several decades the evolution of dating methods, sampling methods, and other methods of analysis has been almost explosive. The increase in knowledge has led to discoveries which were unimaginable only a few decades ago. Dating was revolutionized with the discovery of radioactive isotopes and W.F. Libby's development in the 1940s of the radiocarbon (^{14}C) method of dating organic matter. In science one important discovery frequently leads to another, and Libby's absolute dating method allowed several new scientific fields to materialize.

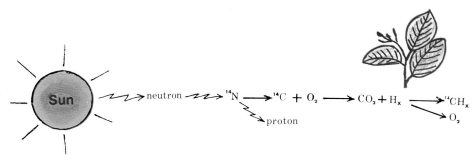

Fig. 1–17. The radiocarbon clock.
The radioactive carbon (^{14}C) isotope is formed in the upper atmosphere by the bombardment of nitrogen (^{14}N) by neutrons emitted from the sun. In contact with oxygen the ^{14}C is immediately oxydized to carbon dioxide ($^{14}CO_2$), which is mixed with the "normal" atmospheric CO_2 where the two other carbon isotopes, the ^{12}C and the ^{13}C, dominate. Finally, the plants use the CO_2 in combination with hydrogen (H) to build their cells where ^{14}C is found in the same amount as in the atmosphere. Since animals live on plants their cells also contain the same amount of ^{14}C. However, as soon as the plant or animal dies, the formation of new cells stops, and the radiocarbon clock starts (see p. 184).

Evolution of micro-paleontology

During the earliest part of the 20th century the study of fossils in sediments was mainly a study of macro-fossils, that is, fossils visible with the naked eye, such as plant leaves, shells and bones. However, these types of fossils were not common in most sediments. Therefore, the recognition of different groups of fossil micro-organisms which often occur in great numbers, especially in marine and lacustrine sediments, led to a revolution in biostratigraphic studies. The micro-organisms are so small that several hundred individuals can frequently be observed through the microscope in a single gram sample. This discovery led to an intensive study of different kinds of micro-organisms, and their evolution and their present-day distribution and living conditions. Again, "the present is the key to the past", and on the basis of the present-day occurrences, conclusions have been drawn about analogous past environments associated with the observed fossil micro-organisms.

Altogether, 8 to 10 different major groups of significant micro-organisms are now generally being studied, many of them primitive planktonic plants and animals that live in the surface water of the oceans (see p. 147). On land, small pollen grains from plants and trees and diatoms from lakes are most important for reconstructing the former vegetation and climate pattern (see Figs. 1–19, 1–21).

Radiometric geological clocks

The discovery of radioactive isotopes, which can be used as precision geological clocks for obtaining the absolute age of rocks and sediments, added a new dimension to geological studies. The radiometric dating of young Quaternary sediments started with the discovery of the radioactive ^{14}C and the associated dating method which W.F. Libby developed (Fig. 1–17). Later several other radioactive isotopes were added to the list for absolute dating (see p. 183). All radioactive isotopes disintegrate at fixed rates, and therefore can be used as geological clocks under special conditions. The three most important of these conditions are: 1. the initial amount of the radioactive isotope contained in the sediment which one wants to date must be known; 2. the geological system in which the isotope occurs must be chemically closed to prevent the addition or subtraction of radioactive material after the "geological clock" began ticking; and 3. the rate of disintegration of the isotope must be known.

The availability of radiometric dates opened the possibility for other forms of exact dating, through cross-calibration with other non-radiometric methods. For instance, the time scale for several processes that take place at a fixed, but formerly unknown, speed can now be determined. An example is the *amino-acid* dating method. Other methods are based on

Fig. 1–18. Sediment core from a lake in Switzerland. The black beds consist of organic mud, and the buff-colored beds represent calcareous marl. The greyish bed at the red knife is a volcanic ash bed deposited during the Laacher Lake eruption in the Eifel district about 11 000 years ago. This ash bed is a typical marker bed which has been observed over a wide area.

the fact that some measurable changes occurred at the same time all over the globe. When these kinds of changes are dated by means of radiometric methods in one part of the world, the dates can be extended to other parts of the globe where corresponding changes are observed. The *paleomagnetic* and the *oxygen-isotope* methods are good examples of this use (see Fig. 1–19). Sedimentary layers which represent short time events and have a wide geographic distribution, such as some volcanic ash beds, may serve as marker layers. When dated in one area, the marker layer can then be used to date beds in other parts of the distribution area (see Fig. 1–18). All of the above methods will be described in Chapter 3, but one of the most broadly useful, the oxygen-isotope method, is introduced in the following section as an example.

Coring in the deep sea

During the early half of the 20th century scientists began to consider the possibility of collecting sediment cores from deep-sea floors where they expected that sedimentation had been continuous throughout the entire 65 million years of the Cenozoic Era. With the development of new coring techniques and the construction of ships specially fitted for work in the deep oceans, this possibility has become a reality. Thousands of long cores from both shallow and deep ocean areas have been recovered and analyzed during the last few decades, and the results are staggering. Studies of the fossil micro-fauna and flora, and of chemical and other sedimentological properties of the sediments, have revealed an almost unbelievably continuous and detailed record of global fluctuations of ocean water conditions and climate conditions during the late Cenozoic time.

The cores recovered from the deep sea revealed zones with cold-water fossils and zones with warm-water fossils. The zones younger than 40 000–50 000 years were dated by means of the radiocarbon method. But what about the age of the older zones where the ^{14}C method cannot be used? Intensive research led to the discovery of another surprising and important dating method, the oxygen-isotope method, which in combination with the paleomagnetic method has been of great importance in dating the zones defined through deep-sea coring (see Fig. 1–19).

The oxygen-isotopic method and stratigraphy

The carbonate in living shells contains the oxygen isotopes of [18]O and [16]O in a ratio which in general depends largely upon the water temperature. Therefore, the ^{18}O/^{16}O ratio in fossil shells usually reflects the contemporary temperatures of the ancient water. However, it was soon discovered that the ^{18}O/^{16}O ratio in shells preserved in cores from the deep sea, where the temperature fluctuations are very small, had a pattern which was mainly dependent upon the size of the glaciers on land and the amount of water which had evaporated from the oceans to form the glaciers. Water molecules with the light [16]O evaporate more easily than molecules with [18]O. Therefore, [18]O is enriched in the oceans and the shells during glaciations. Consequently the ^{18}O/^{16}O ratio graph based on fossil shells records glacials, but also interglacials, and smaller variations in global glacier ice volume. The observed deep-sea

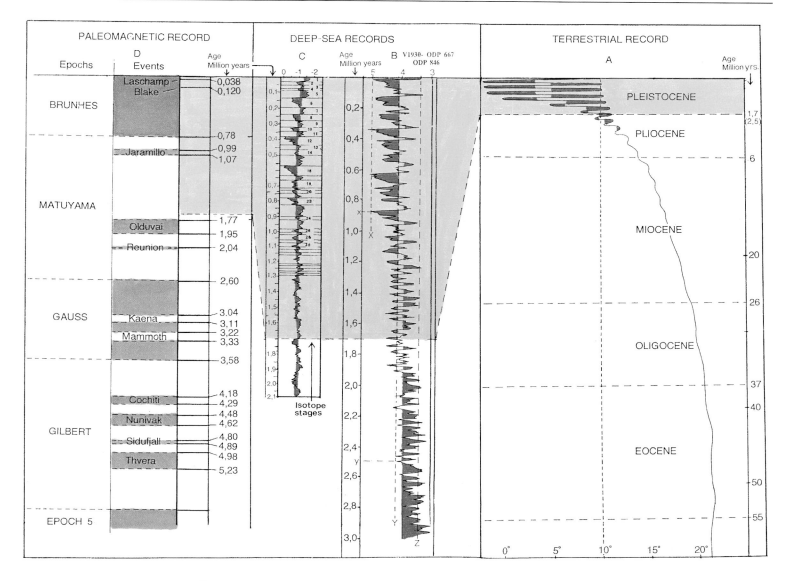

oxygen-isotopic graphs from various global oceans are surprisingly similar, but there are differences in details caused by an overprint of a water-temperature signal, and in some cases by a signal from the meltwater influx resulting from the melting of the large ice sheets.

Figure 1–19 shows two deep-sea $^{18}O/^{16}O$ graphs where the currently recognized isotopic stages or stratigraphic zones are numbered. The sequence starts with stage 1 (the most recent) and continues back in time as indicated. Stage 5e corresponds with the last interglacial (Eem/Sangamon), while stage 6 represents the Saalian/Illinoian Glaciation. The ages presented are partly based on direct radiometric dates from organic remains taken from core samples, and partly on correlation with radiometrically dated events on land. A major factor

Fig. 1–19. Climate fluctuations. A: Climate fluctuations (mean annual temperatures) for the last 50 million years in central Europe (modified from P. Woldstedt, 1954). The time scale has been changed in accordance with the scale in Fig. 1–31. Climate records based on studies of deep-sea cores support the general trend of this graph. However, they indicate that short-time climate and glacier fluctuations occurred through most of the 50 million year period. Yellow: The Quaternary-Pleistocene.

B and C: Records for the last 3.0 and 2.1 million years of deep-sea oxygen-isotope fluctuations, δ^{18}O fluctuations, which correspond with both the global ice-volume fluctuations and global climate fluctuations. (The graphs are modified from N. Shackleton and N. Opdyke, 1976 (C), and Shackleton and others, 1993 (B)).
Blue: glacial periods with large ice volumes. Red: warm periods.
Note the change towards larger ice volume (colder climate) which occurred about 2.5 million years ago (y) and after 0.9 million years ago (x).
Dashed lines: X: Isotope ratio near 0.9 million years ago. Y: Isotope ratio near 2.5 million years ago; Z: Present isotope ratio.

D: The paleomagnetic record for the last 5 million years (modified mainly from N. Shackleton and others, 1990, and F.J. Hilgen, 1991). The record shows the paleomagnetic reversals. Green: normal polarity. White: reversed polarity. The paleomagnetic reversals are very important in dating both marine sediments and various terrestrial sediments such as loess beds.

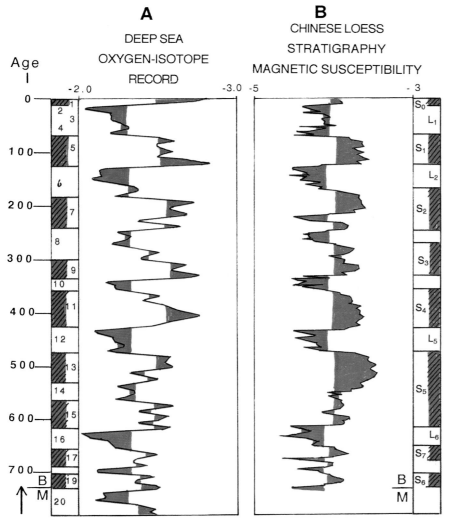

A

B

Age

DEEP SEA
OXYGEN-ISOTOPE
RECORD

CHINESE LOESS
STRATIGRAPHY
MAGNETIC SUSCEPTIBILITY

Thousand years

Fig. 1–20A. Magnetic susceptibility fluctuations in Chinese loess at Xifeng compared with a deep-sea oxygen-isotope record for the last 0.73 million years (modified from G. Kukla, 1987). Red: warm or relatively warm (interglacial or interstadial) phases. Blue: cold or relatively cold (glacial or stadial) phases.

A: SPECMAP-84 age model for the average oxygen-isotope stratigraphy (modified from Imbrie and others, 1984). Numbered isotope stages to the left.

B: A magnetic susceptibility record in loess sections at Xifeng, China.
S_0–S_7: soil horizons in the loess. L_1–L_6: unweathered loess beds.
The magnetic susceptibility is higher in the interglacial/interstadial soils than in the glacial/stadial loess, a difference caused by a higher portion of ferromagnetic minerals in the soils. The exact reason for this difference is not fully understood. B/M: the Brunhes-Matuyama paleomagnetic boundary (see Fig. 1–19).

dating and correlating stratigraphic zones in cores from all over the globe.

The graphs in Fig. 1–19 show many fluctuations between glacial and interglacial periods during the last 2.5 million years (approx.), and also that the amplitudes of the fluctuation have been particularly large during about the last 0.9 million years. This fact indicates that the total ice volume of the world's glaciers was particularly large during the glacials of the last 0.9 million years. Before 2.5 million years ago the total global volume of glaciers was clearly relatively small.

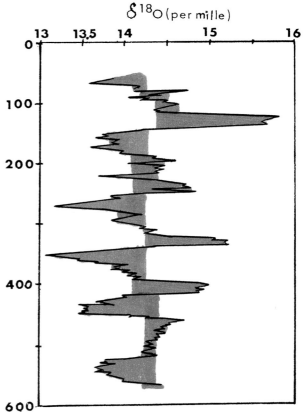

$\delta^{18}O$ (per mille)

AGE (Thousand years)

Fig. 1–20B. $\delta^{18}O$ fluctuations recorded in a 36 cm long core through a calcite unit formed by continuous precipitation from clacite-saturated ground water on the wall of a fissure at Devils Hole, Nevada, USA (modified from Winograd and others, 1992). The graph records the temperature signals of glacials and interglacials back to more than 500 000 years ago. The calcite sediments were Uranium-series dated, and this method is considered to be very accurate. Note how well the graph in general corresponds with the deep-sea graphs, except that the ages of some fluctuations are slightly different.

involved in developing these correlations has been the sequence of paleomagnetic reversals recorded both in the marine cores and in the terrestrial deposits (see Fig. 1–19). Using this method and others, the age of the isotopic stages were fixed, and today they are used in

"Complete" Quaternary climate records on the continents

As already mentioned, the records observed in the Quaternary sediments in various parts of the continents are generally very incomplete as compared with the marine sedimentary record. This is particularly true for the formerly glaciated regions where the glaciers frequently eroded and removed many older sedimentary beds, leaving only fragments of the older record for interpretation. However, there are areas beyond the extent of the ice where fairly continuous records of most of the last 2.5 million years have been recovered. For example, in the Rhine Delta region of Holland, which has experienced fairly continuous subsidence throughout this period, the sedimentary layers have been stacked in a thick sequence. The record presented in Fig. 2–2 was obtained from this region through analyses of cores.

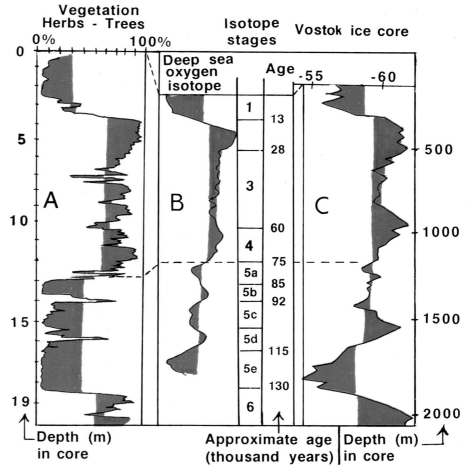

Fig. 1–21. Climate fluctuations during the last 150 000 years (three independent records).

A: Vegetation record from northern France (pollen percentages of herbs and shrubs versus trees). (Modified from G.M. Woillard, 1978.)

B: Deep-sea oxygen-isotope record. (Modified from Martinson et al., 1987.)

C: Ice-core oxygen-isotope record ($\delta^{18}O‰$ from Vostok Station on the South Pole Plateau (see Fig. 1–24). (Modified from J. Jouzel et al., 1987.)
Red: warm, or relatively warm, phases. Blue: cold, or relatively cold, phases. There are many other "climate" graphs based on vegetation changes and on oxygen-isotope fluctuations in cores from the deep sea, ice sheets, and calcite deposits. These show much the same features, in general, as the ones presented here.

Chinese loess stratigraphy

A most complete record of glaciations has been observed recently from the Chinese loess plateau. Some 40 to 50 eolian loess beds have been recorded, and all late Pliocene and Pleistocene glacials and interglacials are probably represented in the stratigraphic sections. Areas of high atmospheric pressure developed over the Tibetan Plateau during the glacial phases, and the resultant increased circulation (including strong katabatic winds) eroded, transported, and deposited the fine-grained, silt-sized rock material, known as loess, in central China. The intervening interglacial phases are recognized as preserved soil horizons formed by weathering in combination with vegetation. Figure 1–20A shows a magnetic susceptibility record observed in loess sections at Xifeng. Note the marked difference in magnetic susceptibility between glacials and interglacials, and how well this record corresponds with the deep-sea oxygen-isotope record. Similar, but not as complete, stratigraphic loess sections have been recorded in central Europe also.

Stratigraphy of calcite deposits

Calcite deposited by calcite-saturated ground water in, for instance, limestone caves and fissures may store surprisingly good and complete records of climate fluctuations through time. The oxygen trapped in the $CaCo_3$ is used to determine the variations in ^{18}O content and thus the temperature fluctuations. Since calcite is well suited for Uranium-series dating, the deposits can usually be dated rather accurately. Figure 1–20B shows a record of oxygen-isotope and temperature fluctuations during most of the past 500 000 years.

Glacier ice cores

Very important information about former climate fluctuations is also stored within the beds of snow and ice in the large ice sheets. Long glacier ice cores, some more than 3000 m long, have been recovered from ice sheets on Greenland, on Baffin Island, and in Antarctica. The youngest parts of many cores display the annually accumulated snow/ice layers rather clearly, and it has been possible to count layers back to more than 30 000 years B.P. Carbon trapped as CO_3 has been used to date core sections younger than 40 000 years, and older sections were generally dated by means of extrapolation methods. All cores have been analyzed for various chemical components which were trapped when the snow was deposited. On the basis of analyses of the oxygen-isotope content, the climatic changes were recorded, and all climate graphs obtained from the cores are strikingly similar. However, the cores analyzed so far cover only about the last 150 000 years. The climate records obtained from the ice cores correspond very well with both the deep-sea records and the other terrestrial records (see Fig. 1–21). Among features which have become strikingly clear through the study of the ice cores are the extremely rapid changes in climate which took place during certain periods.

Glaciations older than the Cenozoic (see Fig. 1–22)

Lithified glacial tills, and various lithified glaciofluvial and periglacial deposits, collectively referred to as "tillites", have been recorded from all of the earth's major continents. They are of different ages. Best known are the Gondwana-series Talchir tillites of Permo-Carboniferous age. About 280 million years ago India and all the southern hemisphere continents, Australia, Africa, South America, and Antarctica, were joined in the Gondwana supercontinent, which was variously covered by large ice sheets through time. Other well-authenticated tillites of earlier ages lie, for example, in the exposed rocks of the Sahara Desert of Africa. These consist of two separate units of tillite, one about 700 million years old (Precambrian) and the other about 450 million years old (Ordovician). Other 700 million year old tillites have been recorded, for instance, in Canada and Scandinavia. Such tillites are considered to represent large-scale glaciations resulting from global cooling. The cold climate which resulted in large expansions of the glaciers during the last 2.5 million years represents the most recent of such major cold periods.

The Cenozoic Era cooling of the climate, 50 million – 2.5 million years ago; glacier fluctuations in the polar regions

The Cenozoic climate record was originally derived from observations on the continents. For instance, records from central Europe demonstrated that a gradual cooling occurred from about 40 million to about 2 million years ago, when a time of more dramatic climate fluctuations started (see Fig. 1–19). However, these records were incomplete and revealed little detailed information about these climate fluctuations. With the addition of information derived from the deep-sea cores during the last few decades, new and much more detailed interpretations have been made possible. Many cores with a continuous or nearly continuous sediment record of the entire Cenozoic Era have been recovered from all of the world's major oceans, and have allowed the documentation of a very interesting story. Specialists in several disciplines have analyzed the cores. Stratigraphic and climate zonations and graphs have been made based on analyses of, for example, fossil foraminifera, diatoms, radiolaria, coccoliths, oxygen isotopes, organic carbon, biosilicates and ice-rafted rock material. Mostly planktonic species are represented, but some benthic organisms have also been studied. The combined results reveal a rather consistent picture where climate interpretations based on, for instance, foram studies agree with those based on studies of other organisms, oxygen isotopes and organic carbon.

The general climatic trend observed on land has been verified in the deep-sea records, which show a general drop in temperature from the middle Eocene, 45–50 million years ago, to the late Cenozoic Ice Age period, which "started" about 2.5 million years ago (see Fig. 1–19). The early and middle Eocene climate was very warm, and no glaciers existed even in the high-latitude polar regions. The latitudinal temperature gradient, the drop in temperature from the poles towards the equator, was low, as was the vertical temperature gradient of the oceans. The cooling of the climate following the middle Eocene time was accompanied by an increased oceanographic circulation. Both the latitudinal and the vertical temperature gradients gradually increased, and they seem to have increased faster (in steps) during certain intervals. A first prominent step occurred near the Eocene-Oligocene transition, and it corresponds with the first known formation of glaciers in Antarctica. Other steps seem to have occurred about 15, 10, 5, 2.5 and 0.9 million years ago. Causes for the steps have been very actively debated. However, an obviously important factor was the drastic reorganization of the ocean current system during some of these periods. Oceanic gateways were opened and closed as a result of the plate-tectonic-induced migration of the continents, and marine sills/thresholds were lowered and raised. For instance, the lowering of the sill across the Atlantic Ocean between Greenland and Scotland was of immense importance for the flow of the cold deep/intermediate water current between the North and the South Atlantic.

Glaciers formed much later in the Arctic than in Antarctica. The approximately 6 million year old glacial deposits in Alaska represent an old Arctic glaciation, but it has been suggested that local Alpine glaciers formed as early as 9 million years ago in some Arctic mountains. This corresponds well with observations in cores from the Arctic Ocean and the North Atlantic, where ice-rafted clasts have been found in marine sediments as old as about 10 million years.

Former forests in Antarctica and in the Arctic

Although the general climatic trend during the last 50 million years was towards a gradually colder climate, there were also times of relative

Fig. 1–22. Lithified glacial tills (tillites) and glacially striated bedrock which are evidence of former glaciations, 700, 450, 280 and 2 to 5 (?) million years old.

A: Tillite resting on glacially striated rock surface at Bigganjargga in northern Norway. The tillite is about 700 million years old (see Fig. 1–15A).

B: A large glacially striated erratic in a tillite about 700 million years old in the Mauritanian Sahara Desert, Africa.

C: The Talchir Tillite in central India is of Permo-Carboniferous age, about 270–290 million years old. Note that the larger erratics have a glacial striation parallel with the knife.

D: The Sirius Formation tillite at Beardmore Glacier in Antarctica, about 5° from the South Pole. Beardmore Glacier in the background. Fossil twigs of Nothofagus *(x) were found in lake deposits (B) sandwiched between two tillite beds (A+B). Fossils within the beds suggest a Pliocene age (2 to 5 million years) for the tillite, but other evidence suggests a Miocene age. At that time the valleys in the Transantarctic Mountains near the South Pole were partly forested and valley glaciers pushed into the lakes.*

E: The Ordovician and the Permo-Carboniferous ice sheets existed at a time when the southern hemisphere continents were joined in a supercontinent.

Fig. 1–23. The climate, vegetation, animal life, and landscape have changed drastically during the last 20 million years, as illustrated by three scenes from the Lucerne area in Switzerland.

A. The Lucerne area 20 million years ago, when the climate was very warm. (Painting from "Glacier Garten" collection.)

B. The same area 20 000 years ago, during the last maximum glaciation. (Painting from "Glacier Garten" collection.)

C. The same area today, with the Alps in the background, Vierwaldstätter Lake in the middle, and Lucerne City in the foreground. The lake is one of the many beautiful glacially sculptured lakes at the foothills of the Alps, and the entire alpine landscape is glacially sculptured. Most of the farmland adjacent to the city lies on glacial or glacial-related deposit. (Photo by Swiss Air.)

warmth. Such relatively warm periods were recorded in Antarctica during middle Miocene time, about 17 to 15 million years ago, in late Miocene time, and in Pliocene time, 5 to 4.5 million years ago. The Pliocene (or Miocene?) Sirius Formation tillites in the Transantarctic Mountains contain lake beds with twigs and leaves of *Nothofagus* trees found in an area about 5° from the South Pole. This shows that the valleys in the Transantarctic Mountains at that time were more or less forested close to the South Pole, and that valley glaciers pushed into lakes on the valley floors (see Fig. 1–22).

A relatively warm climate existed in the Arctic during Paleocene and Eocene times (67–37 million years ago), and deciduous and coniferous forests covered much of the Arctic coasts. However, the marked global climatic cooling near the Eocene-Oligocene transition probably affected the vegetation in the Arctic also, and from then on the forest composition gradually changed to become a Taiga and forested tundra in Pliocene time. The open tundra which covers the Arctic coasts today was established as late as near the end of Pliocene time.

Figure 1–23 illustrates the warm conditions in central Europe in Miocene time, about 20 million years ago.

Climatic changes after 2.5 million years ago; formation of mid-latitude ice sheets; the "true" late Cenozoic Ice Age

Both the deep-sea oxygen-isotope record and analyses of ice-rafted clasts in the glaciomarine deposits in the Arctic and Antarctica show that a cooling and a considerable expansion of glaciers took place about 2.5 million years ago. Terrestrial glacial deposits of about this age have also been recorded, for example, from Iceland, the Midwest region of USA, and South America, showing that mid-latitude ice sheets had formed. From then on, throughout the late Pliocene and the Pleistocene, the mid-latitude glaciers existed and fluctuated. The amplitudes of the fluctuations increased after 0.9 million years ago, when the largest mid-latitude ice sheets formed, supposedly in response to the 100 000 year Milankovitch cycles (see Fig. 1–26).

During the coldest phases, the true ice ages or glacials, the ice sheets expanded over large areas of North America and northern Europe, and much of the North Atlantic Ocean was covered by sea ice (pack ice) and ice shelves. During the intermediate warmer phases, the interglacials, the climate and glacier conditions were about the same as they are today. These glacier fluctuations are recorded both in the oxygen-isotope graphs and in the terrestrial stratigraphy, which will be discussed later.

What was the cause of the long-term and the short-term climate and glacier fluctuations?

The ultimate cause for the long-term climate and glacier fluctuations still remains much of a mystery. Over time several hypotheses have been proposed, such as: 1. long-term changes in the interstellar position of the earth; 2. variations in solar activity; 3. variations in atmospheric carbon dioxide; 4. changes in ocean circulation, caused by the drift of the continents and closing or opening of ocean gateways; and 5. the changing altitude of the major mountain chains, resulting in major changes in the atmospheric circulation.

The cause for shorter-term climate and glacier fluctuations during the late Pliocene and the Pleistocene has also been actively debated. The known fluctuations are shown in Fig. 1–19. Some of the coldest periods, which represent the glacial phases, had mean annual temperatures in the order of 10–20°C lower than today in areas beyond the ice sheets in parts of Europe and USA. In parts of the subtropics and continental tropics the temperatures were about 4–7°C lower, and along the marine equator, they were only slightly lower, or no lower, than today. The warmest phases between the cold glacial phases, the interglacials, had temperatures analogous to the present, although in the warmest parts of some interglacials the mean summer temperatures were in the order of 2°C warmer than today in much of Europe and North America.

Ever since it was discovered that the climate had fluctuated significantly, causing alternating glacials and interglacials during the last 2.5 million years, scientists have speculated about the cause for the fluctuations. Numerous theories have been proposed, of which some of the best known suggest the cause to be changes related to: 1. astronomic factors; 2. sunspot activity; 3. ocean currents; 4. atmospheric composition; 5. volcanic dust or dust from disintegrated meteorites in the atmosphere; 6. surges of the Antarctic Ice Sheet; 7. rising and falling of parts of the earth's crust; 8. solar-terrestrial magnetic coupling; 9. fluctuations of the upper atmospheric jet-streams; 10. asteroid impacts; and 11. interaction between ocean currents and atmospheric circulation. In addition, the recently formulated "snowblitz" theory suggests that a series of harsh winters and cool summers started the glaciations.

Objections have been presented against all of these theories, and many of them have been rejected outright as very unlikely. A requirement that most of them fail to fulfill is an explanation of the fairly regular pattern of glacials and interglacials, as shown in the oxygen-isotope graphs. The theory that seems to fit this pattern best is the astronomic Milankovitch theory. However, two serious objections have been raised against this theory. First of all the theory requires glacial and climate fluctuations in the northern hemisphere to be out-of-phase with the fluctuations in the southern hemisphere. Most field observations indicate that this is probably not the case, and that in fact the fluctuations are in phase in both hemispheres. Second, the calculated changes in energy-input to the earth's atmosphere and the earth's surface caused by the Milankovitch-predicted changes are too minor as compared to the magnitude of the actual changes. For these and other reasons, many scientists rejected the Milankovitch theory. However, new research carried out during the recent decades reveal evidence of the Milankovitch forcing signals in the deep-sea oxygen-isotope graphs, and even in the oxygen-isotope graphs from analyses of ice cores from the large existing ice sheets and in the magnetic susceptibility graphs from the Chinese loess deposits. Therefore, most scientists now seem to accept that the astronomic (Milankovitch) signals represent the most likely triggering mechanism for the late Cenozoic climate and glacier fluctuations (Fig. 1–26). Today much research is focussed on finding the mechanism by which the relatively weak astronomic signals are ampli-

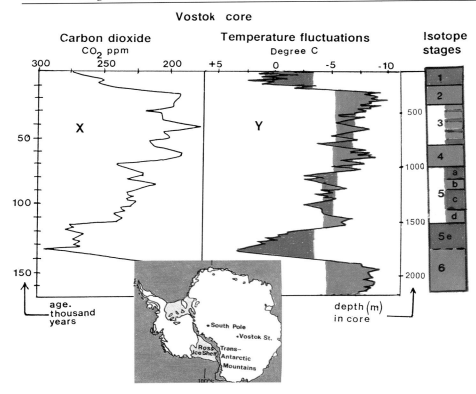

Fig. 1–24. *Fluctuations in atmospheric CO_2 and temperatures at Vostok Station on the Polar Plateau, Antarctica, based on observed chemical components in the collected ice core (CO_2: modified from J.M. Barnola and others, 1987; temperature: modified from J. Jouzel and others, 1987). Suggested correlation with the deep-sea isotope stages is added. Red: warm, or relatively warm. Blue: cold, or relatively cold. Note how closely the CO_2 graph corresponds with the temperature graph. This has led to a discussion about which is the driving force, the changes in CO_2 or the changes in temperature. Apparently the temperature changes were slightly ahead of the CO_2 changes, but most scientists seem to agree that there is a close interaction between the atmospheric temperature and CO_2 content.*

Fig. 1–25. *The "conveyor belt" which supposedly shows the main oceanic heat flow, with a deep, cold, and saline water current (green), and a warm, less saline surface current (yellow). (Modified from W.S. Broeker, 1987.) The conveyor belt represents a gigantic, global, oceanic heat-flow system. It was established in late Cenozoic time, and changes in this system, including changes in direction and strength of the individual surface currents, participated in the forcing of ice age climate fluctuations.*

The diagram of the conveyor belt shows a simplified and generalized system, and in detail the current system is more complex. For example, the surface currents, which are usually driven by the prevailing winds, have a much more complex pattern. Some of the surface currents in the North Atlantic, which are of utmost importance for the climate in western Europe, Greenland, and eastern North America, are plotted on the diagram. In red: the Gulf Stream (G), which heats the coasts of western Europe. It was diverted southwards and never reached these coasts during the coldest parts of the glaciation.

In blue: the cold East Greenland (EG) and Labrador (L) currents, which are responsible for the cooling of the adjacent coasts.

The importance of these currents, in combination with the atmospheric wind systems, is illustrated by the fact that Greenland today is covered by a large ice sheet on the same latitude as northern Scandinavia, which is essentially unglaciated and has a significantly warmer climate.

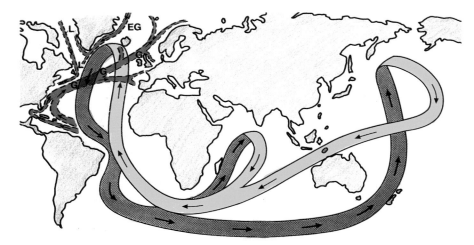

fied, and to explain why the climate forcing for the mid-latitude northern hemisphere seems to govern the climate pattern in the southern hemisphere also.

Clearly several factors must be involved in the amplifying process. Many of the factors which have been suggested as causes for the glaciations must have been involved; among them are all the changes in atmospheric circulation and atmospheric composition. Today the carbon dioxide in the atmosphere absorbs about 15% of the energy that is reflected from the earth's surface. However, during glaciations, the production of CO_2 and the atmospheric content of CO_2 was much smaller than today (Fig. 1–24), and it has been suggested that the rise in atmospheric CO_2 after the glaciations was responsible for about one-third of the rise in air temperature. Another example relates to changes in the strength and the paths of ocean currents, both the important cold deep-water bottom currents and the surficial currents (Fig. 1–25). For instance, the warm-water Gulf Stream, which today flows northwards in the North Atlantic and heats the west coasts of northwestern Europe, was deflected southwards along the coasts of Portugal and West Africa, and never reached the northern areas during the times of the last maximum glaciation. The extent of the snow cover on the earth's surface was important, too, since white snow generally reflects more than 70% of the incoming solar radiation, in contrast, for example, to 10–20% from some forests. Therefore, the expansion of snow-covered areas during the glaciations had a considerable cooling effect. Other factors can be added, and they probably all had some effect in creating the ice age climate. However, the triggering factors for the Quaternary climatic changes are generally attributed to the astronomic factors, although this theory has been questioned by some scientists (see next chapter).

The astronomic theory (Fig. 1–26)

Several prominent scientists were responsible for the evolution of the astronomic theory. The French mathematician J. Adhemar in 1842, the Scottish physical scientist J. Croll in 1864, and the Serbian engineer and mathematician M. Milankovitch, in 1911 and 1930, all provided pieces of the solution to this puzzle. Croll

reported that the cold peak of the last glaciation was about 80 000 years ago, but according to Milankovitch, who performed more elaborate calculations, the last cold peak occurred about 25 000 years ago, which is very close to the estimates from field observations.

Three main astronomic factors influence the insolation received on the earth:

l. *Eccentricity*. The earth's orbit around the sun is slightly elliptical, and the elongation of the ellipse (the eccentricity) changes in 100 000 year cycles.

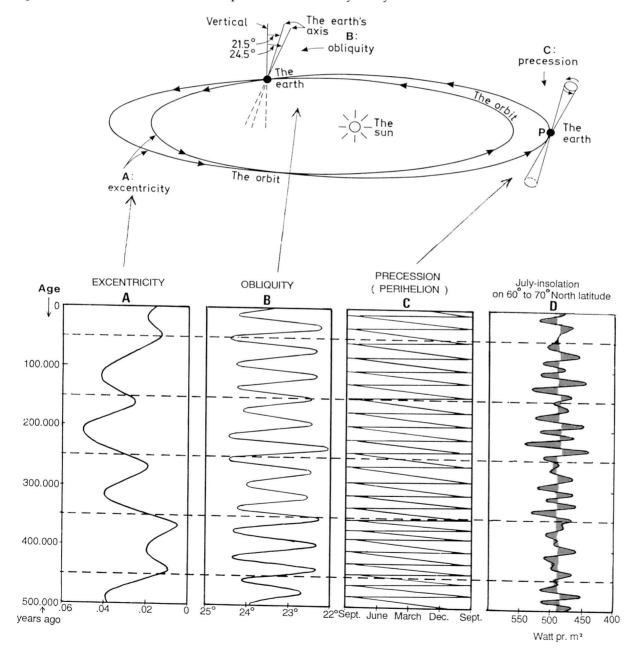

Fig. 1–26A.
The astronomic factors, also called Milankovitch factors, which are believed to determine the main climatic pattern during the late Cenozoic.

A: The eccentricity of the earth's orbit varies in 100 000 year cycles.

B: The obliquity: The tilt of the earth's axis relative to the orbital plain fluctuates in 41 000 year cycles.

C: The precession fluctuates in 23 000/19 000 year cycles, resulting from the wobbling of the earth's axis. P: Perihelion is the point on the earth's orbit which is closest to the sun.

Fig. 1–26B.
Calculated fluctuations of the Milankovitch factors during the last 500 000 years, and the resulting fluctuations of insolation to the earth on the 60° to 70° North latitudes.

A: Eccentricity. B: Obliquity. C: Precession (when Perihelion is closest to the sun). D: The fluctuation of insolation to the earth on the 60° to 70° North latitudes, as a result of the fluctuations of all Milankovitch factors combined. Red: warm. Blue: cold periods. (Modified from C. Covey, 1984.)

A

B

C

Fig. 1–27. The regional continental and marine bio-zones in Europe, the North Atlantic, and eastern North America, generally outlined. All zone boundaries are transitional. Note the east-west trend of the boundaries, except near the coasts, where they are diverted in a northerly or southerly direction by, respectively, the Gulf Stream (G) and the East Greenland and Labrador currents (E and L). Illustrations of some typical terrestrial forest zones are added.

A – D Arctic and Subarctic vegetation:
A: Arctic, open tundra (white).
B: Arctic, semi-open tundra/parkland with patches of birch forest (white).
C: Subarctic, semi-open birch forest (dark blue).
D: Subarctic, transition from birch to spruce forest (dark blue).

E-G Boreal and deciduous broad-leaf forest:
E: Transition from birch to a more typical Boreal spruce-pine forest (green).
F: Boreal pine-spruce forest (green).
G: Broad-leaf/mixed oak forest (green).

The Arctic and Subarctic vegetation (A to D) dominated most of Europe during the coldest phases of the late Cenozoic ice ages.

D

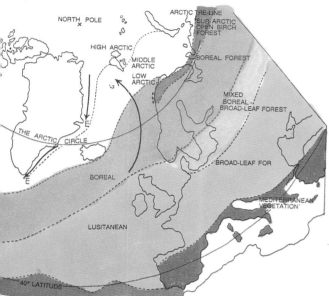

2. *Obliquity*. The earth's axis has today a tilt of about 23.5° to the orbital plane, and that tilt varies between about 24.5° and 22.5° in 41 000 year periods.

3. *Precession*. The earth's axis wobbles in space and describes a circle which results in about 19 000 to 23 000 year cycles.

As previously mentioned, the imprints of the 100 000 year cycles, the 41 000 year cycles, and the 19 000 to 23 000 year cycles are recognized in the deep-sea oxygen-isotope graphs. In additon, the 100 000 year cycles seem to best account for the major pulses of glaciations during the last 0.9 million years. However, the 100 000 year (eccentricity) insolation pulses are rather weak, and it has been a problem to explain how and why their effect could be so strongly amplified during the last 0.9 million years. Some scientists even question the importance of the 100 000 year astronomic cycles, and the graph for the temperature fluctuations at Devils Hole has been used as an argument in this connection (see Fig. 1–20B). This graph resembles very much the deep-sea graphs, but the obtained ages are not as well in accordance with the astronomic signals as the ages of the fluctuations in the deep-sea graphs. The Uranium-series dating method used for the Devils Hole sediments is considered very accurate, but good arguments have been presented in favor of a solution/sedimentation mechanism

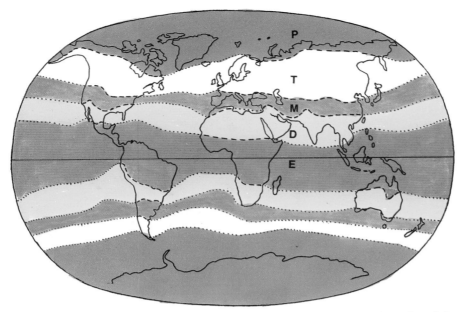

Fig. 1–28. *The world's climate/temperature zones. Note the east-west orientation of the zones. All zone boundaries were moved towards the equator during the ice ages.*

P: *Polar/subpolar cold zone. Mean temperature for the warmest month: less than 10°C.*

T: *Temperate to cool-temperate zone. Mean summer temperature: from 15–22°C.*

M: *Mediterranean warm zone. Mean of coldest month: more than 0°C.*

D: *Tropical, very warm, and generally very dry zone. Mean of coldest month: more than 10°C and less than 20°C.*

E: *Equatorial hot and moist zone. Mean of all months: more than 20°C.*

(*Modified from various sources.*)

Fig. 1–29. *Little Ice Age (1500–1930): marginal moraines in front of Isfall Cirque Glacier near Tarfala in the Kebnekaise Mountains, Sweden. (The outermost moraine on the right side is reckoned to be 2500 years old.) The moraines were probably deposited during small climate fluctuations. Moraines of this kind, which were deposited during Weichselian/Wisconsin glacier oscillations, are used to record former glacial and climatic conditions (see Fig. 1–30). (Photo by Nils Haakonsen.)*

at Devils Hole, which could result in a small displacement of the calcite-temperature cycles relative to the true temperature cycles. Therefore, most scientists still believe that the astronomic forcing is involved in the triggering of the major late Cenozoic climate changes and glaciations. But there are definitely still many unsolved problems related to the mechanism behind the climate changes. For instance, what causes the rapid and dramatic temperature changes of short duration? Some of these changes are in accordance with the astronomic forcing, but others are not, such as the rapid cooling which led to the Younger Dryas cold phase. It has been suggested that these kind of changes may occur when the atmosphere reaches a state of instability which leads to drastic climatic changes. Instability in the atmospheric carbon-dioxide (CO_2) system has been mentioned as a candidate for such changes.

Cyclic climatic changes of shorter duration

Former glacier fluctuations recorded by means of end moraines in front of existing glaciers, together with annual variations in tree-ring growth, variations in annual snow/ice layers in ice cores from the ice sheets, and variations in fauna and flora recorded in stratigraphic sections, all demonstrate that the climate experienced fluctuations of short duration also. Many scientists claim that these fluctuations are cyclic. Periods of about 2500, 1500, 1000, 800, 650, 250, 200, 33 and 11 years have been proposed. For instance, the ice-core records from Greenland show the presence of 2500 year cycles (the Dansgaard-Oesgher cycles), while evidence has been presented for 2500 year cyclic glacier fluctuations in Scandinavia. However, the problem still remains unsolved if these and shorter-duration fluctuations are worldwide and in phase. The Little Ice Age cold phase following A.D. 1500 seems to have affected broad areas in both hemispheres, as did the warmer phase which caused the prominent glacial retreat after A.D. 1930. Therefore, several scientists seem to favor climatic models with worldwide, in-phase fluctuations, but they often disagree on which of the proposed fluctuations are of that kind. They also disagree on their causes. The following are some of the suggested candidates:

(a) variations in sunspot activity;
(b) changes in ocean surface currents;
(c) instability in the atmospheric CO_2 system;
(d) surges of the Canadian Ice Sheet.

Past climatic changes; how can they be recognized?

Several methods by which former climatic changes can be recognized have been described in previous sections. Therefore, only some supplementary methods and a summary of some main principles used in this research will be presented.

The climatic record of the past is stored in the sediments and the rocks which were formed during the past. To read that record requires a basic knowledge about how such sediments are formed today. Again, "the present is the key to the past", and climatic interpretation rests on an intensive study of present-day climatic processes and their results in the various climatic zones of the earth. The present-day distribution of the various plant and animal species and of many physical processes are, to a considerable extent, limited by climate factors such as temperature and precipitation. In the ocean the temperature, salinity, and the supply of nutrients are the three significant factors.

Figure 1–28 presents an outline of the earth's temperature zones, and Fig. 1–27 outlines the major present-day bio-zones in northern Europe, eastern North America, and the North Atlantic Ocean. The zones have, in general, a nearly east-west orientation, although the orientation is somewhat diverted along the ocean margins. Temperature is a main factor in determining the location of the boundaries, and with changes in the temperature, the zone boundaries will move southwards during a cooling and northwards during a warming. Such changes are reflected in the corresponding sediments preserved in stratigraphic sections. For instance, the ice age boundary between the Subarctic and the Arctic zones was located in southern France while it is located in northernmost Scandinavia today, and the Arctic treeline, which now lies near the Arctic coast in northernmost Scandinavia, was located near the north coast of the Mediterranean Sea.

A corresponding vertical climate zonation related primarily to temperature exists on the

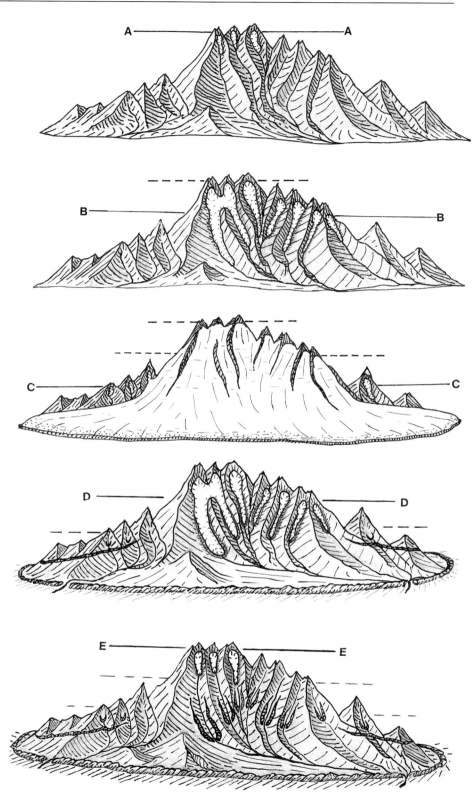

Fig. 1–30. A glacial cycle in an alpine mountain district. During the early part of a glaciation, the glaciation limit is lowered from A-A via B-B to C-C, and during the later part it rises via D-D to E-E. Moraines deposited along the ice margins during the phases C-C, D-D and E-E can be used to reconstruct the corresponding glacier conditions and, in that way, record the corresponding glaciation limit and equilibrium line which are related to the former climate (see Glossary).

slopes of high mountains also. Even some mountains located in the tropical or subtropical zones do rise to altitudes where the climate resembles that in the Arctic zone. This high zone on a mountain is called the Alpine zone, and the zone below, which in many ways corresponds with the Subarctic zone, is called the Subalpine zone regardless of the altitude.

The distribution of moisture and precipitation frequently has a more complicated pattern, depending upon wind direction, topography and distance from open ocean water. For instance, the west coast of northern Europe has a typical oceanic climate with high precipitation, but the precipitation, in general, decreases eastwards, where areas become very dry and continental in parts of Russia. However, during the ice ages the climate in much of western Europe was, at times, about as dry and continental as it is in the east today.

Changes in the altitude of the *snowline* and *glaciation* limit also result from climatic changes, and evidence of such past changes can be detected, especially in Alpine regions (Fig. 1–29). There, Alpine glaciers form on mountains which rise above a certain level, usually called the glaciation limit or glaciation threshold, while mountains with summits below this level are not glaciated. During the ice ages the glaciation limit lay more than 1000 m lower than today in many regions, as evidenced by many Alpine ice age end moraines on low-lying mountains (see Fig. 1–30). The altitude of the glaciation limit and of the snowline depends primarily on a combination of summer temperature and winter precipitation (see p. 105), and frequently the glaciers seem to be more sensitive to climate changes than both fauna and flora.

The oxygen-isotope thermometer

The atmospheric oxygen is composed of two isotopes, ^{16}O and ^{18}O, and the ratio in which the two isotopes enter chemical compounds is temperature-dependent. Therefore, oxygen-isotope ratios observed in, for example, calcareous shells, speleothems, and snow which were formed at various times can be used to interpret the temperatures at the times of their formation. Observations of this kind have been made on cores from lakes, glaciers, and lime-stone (calcite) deposits, such as speleothems in caves. However, a problem with this kind of study has been to distinguish the temperature signal from other noise signals, such as the ice-volume signal (p. 21) and a salinity signal for compounds formed in water.

The geological time scale

The geological time scale in Fig. 1–31 presents both geological time units and some of the more pertinent geological events. Note that the *Quaternary Period* is by far the shortest of all geological periods. However, representing approximately the last 1.7 million years, the deposits from this period are generally well preserved and well exposed at or near the earth's surface. In addition, methods available to date these younger deposits are generally more accurate than the methods used to date the older deposits and rocks. Therefore, information about the Quaternary Period is not only more voluminous, but also more detailed than for earlier geological times.

The beginning of the Quaternary Period, which is equated with the base of the Pleistocene Epoch, has been a problem to define. It was generally placed in the stratigraphic columns from different localities where the first distinct ice age cooling was recorded, as indicated by the presence of the first fossils with cold-environment affinities. However, the beginning of the Quaternary, defined in this manner, varies widely in age from one area to the other. To obtain a time-fixed boundary it was therefore recommended in 1948 by the International Geological Congress to define the base of the Quaternary and the Pleistocene where the cool-water molluscs, the so-called "Nordic guests", first appear in the marine stratigraphic sequence in Italy. However, this decision did not stop the discussion about the stratigraphic position of the base of the Quaternary. Therefore, a committee was formed to provide a better solution to the problem. After a decade of intensive work the committee recommended that the base of the Sicilian beds in the stratigraphy of southern Italy should define the start of the Pleistocene and the Quaternary, and this was accepted by the International Union of Geological Sciences in 1986.

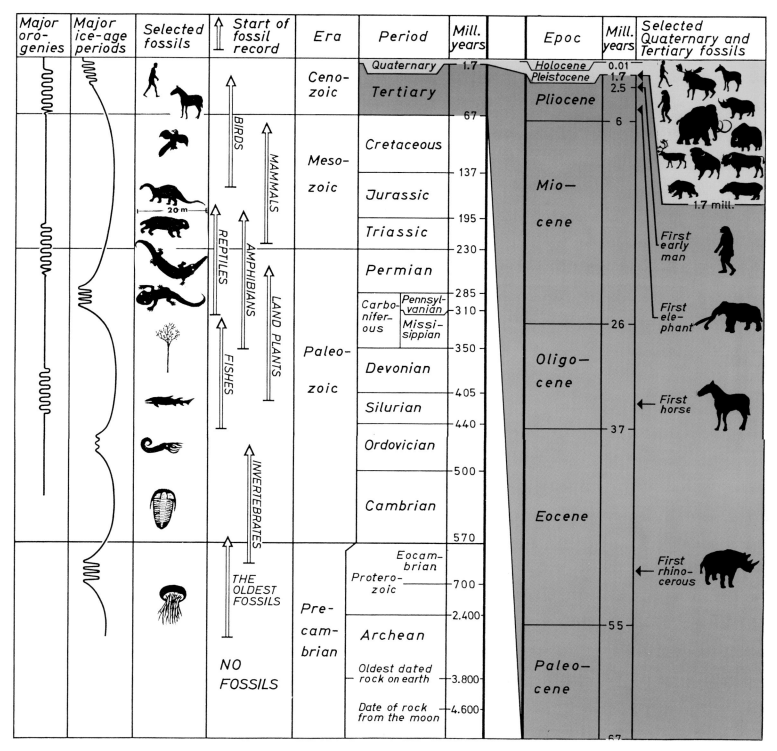

Major oro-genies	Major ice-age periods	Selected fossils	Start of fossil record	Era	Period	Mill. years	Epoc	Mill. years	Selected Quaternary and Tertiary fossils
				Ceno-zoic	Quaternary	1.7	Holocene	0.01	
					Tertiary		Pleistocene	1.7	
								2.5	
			BIRDS			67	Pliocene		
					Cretaceous			6	
			MAMMALS	Meso-zoic		137	Mio-cene		
					Jurassic				First early man
		20 m				195			
			REPTILES		Triassic				
			AMPHIBIANS			230			
					Permian			26	First ele-phant
			LAND PLANTS			285			
					Carbo-nifer-ous / Pennsyl-vanian	310	Oligo-cene		
			FISHES	Paleo-zoic	Missi-sippian	350			First horse
					Devonian			37	
					Silurian	405			
						440	Eocene		
			INVERTEBRATES		Ordovician	500			
					Cambrian				
						570			First rhino-cerous
			THE OLDEST FOSSILS	Pre-cam-brian	Eocam-brian				
					Protero-zoic	700		55	
						2.400			
			NO FOSSILS		Archean		Paleo-cene		
					Oldest dated rock on earth	3.800			
					Date of rock from the moon	4.600		67	

1.7 mill.

Fig. 1–31. The geological time chart. Some of the most important and interesting geological events are emphasized. The selection of a few of the better-known Quaternary/Pleistocene mammals are shown. They include both cold- and warm-climate species. The presented ages of the zone boundaries may vary somewhat in different publications.

The newly defined boundary lies just above the Olduvaian paleomagnetic zone, which has an upper boundary radiometrically defined at about 1.77 million years ago. Therefore, the age of the lower boundary for the Quaternary is also defined at about 1.7 million years ago. However, many Quaternary scientists were none too happy with this definition. They were used to considering the Quaternary as more or less identical with the time of the late Cenozoic mid-latitude glaciations, which covers approximately the last 2.5 million years (see p. 35). In addition, the oldest remains of humans seem to be about 2.5 million years old, and many geologists have considered the Quaternary to be the "Era of humans". Therefore, there are still attempts made to have this boundary redefined.

Subdivision of the Quaternary Period and the Pleistocene Epoch

The Quaternary Period is subdivided into a Pleistocene Epoch and a Holocene Epoch, of which the Holocene represents the last 10 000 years. The boundary, set at about 10 000 years ago, was determined by the marked change in climate from cold glacial to warm interglacial which occurred at that time over broad areas of the world.

The Pleistocene Epoch is generally subdivided into an early Pleistocene (1.6 to 0.7 million years ago), a middle Pleistocene (0.7 to 0.13 million years ago), and a late Pleistocene, comprising the last interglacial-glacial cycle (130 000 to 10 000 years ago).

Fig. 1–32. The skjaergaard within the "Skjærgårdspark" on the coast of southern Norway consists of numerous skerries and small, low islands sculptured by ice age glaciers. This kind of landscape is typical for many formerly glaciated coastal districts, such as many coasts in eastern Fennoscandia, including the famous Stockholm skjaergaard and the finnish skjaergaard. They are the result of ice-sheet erosion on flat or low undulating terrain where no deep valleys directed the flow of thick and fast moving ice streams. The skjaergaard landscape, the fjord landscape, the lake landscape and the alpine landscape represent four major types of glacial erosion landscapes.

Fig. 1–33. Lake Gjende (984 m above sea level) and Lake Bess-vatn (1373 m above sea level) within Jotunheimen Mountain in central Norway are two of the many beautiful mountain lakes which occupy rock basins sculptured by ice age glaciers.

Chapter 2
ICE AGE HISTORY, 2.5 MILLION YEARS AGO TO THE PRESENT

with Emphasis on the Land Areas of North America and Northern Europe over the Last 130 000 Years, the Last Interglacial-Glacial Cycle

The dramatic history of the formation and dissipation of the large late-Cenozoic ice sheets has become known through field and laboratory studies carried out by numerous scientists who have been trained along the research lines to be described in Chapter 3.

Much of the presentation in this chapter will be accomplished by the use of map reconstructions for different past time intervals, so-called paleogeographic maps. These maps present very generalized pictures of important landscape features, rather than focussing on extreme details. Readers should notice also that the presented ages for periods younger than 40 000 years are generally radiocarbon ages, while older ages (40 000–200 000 years) are generally obtained by means of Uranium-series dates. However, the order of magnitude of these have usually been verified by dates obtained by the various other dating methods presented on p. 148.

Glacials and interglacials, 2.5 million – 130 000 years ago

Glacials

Glacial sediments in drift sheets deposited by large mid-latitude glaciers have been observed on several continents. Frequently one drift sheet is stacked on top of another, with varying geographic overlap, and as many as four main sheets have generally been recognized. Sandwiched between the drift sheets are often beds with warm-climate interglacial deposits (Fig. 2–1). However, seldom have more than two major drift sheets been recognized in a single locality. The four sheets were, for a long time, believed by most scientists to represent all of the late Cenozoic glaciations, and it was usually suggested that they all occurred during the last 1 million years. However, deep-sea oxygen-isotope graphs (Fig. 1–19) and some

The Alps		North Europe		North America	
Glaciation	Intergl.	Glaciation	Intergl.	Glaciation	Intergl.
Wurm		Weichsel		Wisconsin	
	R-W		Eem		Sangamon
Riss		Saale*		Illinoian	
	M-R		Holstein		Yarmouth
Mindel		Elster		Kansan	
	G-M		Cromer		Aftonian
Günz		Menap		Nebraskan	

Fig. 2–1. The traditional chart of glacials and interglacials recorded on land in northern Europe, North America and the Alps. The table shows the classical correlation which was based on the assumption that there were only four major glaciations. Today parts of this correlation are questioned. Various names are used for the glacials and interglacials in different countries (see "glaciations" in the Glossary). x: The Saale Glaciation is divided into a Drenthe Stadial (oldest) and a Warthe Stadial, with a Trene Interstadial between.
Observations in the Alps suggest that there was at least one pre-Günz glaciation, the Donau Glaciation, and the deep-sea stratigraphy indicates that there must have been several more global glaciations in addition. According to the stratigraphic code the ending "ian" (Weichselian etc.) should be added for the chonostratigraphic stages which correspond to the glaciations and the interglacials (see "stratigraphy" in the Glossary). This rule has frequently not been followed for the North American stages, however.

terrestrial records (Fig. 2–2) show that there must have been many more glaciations distributed over a considerably longer time span. They show, for example, that shortly after 2.5 million years ago there was a time of relatively extensive glaciation, and recent studies on the continents have revealed that glacial deposits of about that age exist in several mid-latitude areas.

New dates obtained by modern radiometric dating techniques indicate that terrestrial deposits formerly assigned to the "original" four glaciations probably represent the deposits of several more glacial periods. The glacial deposits identified at some of the original-type

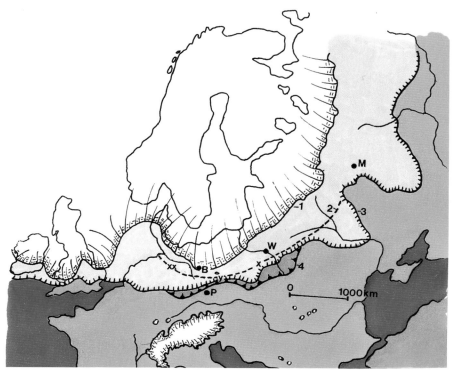

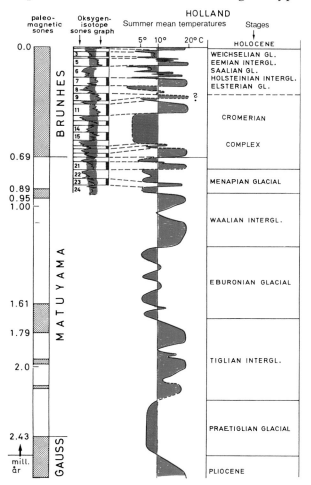

Fig. 2–2. Stratigraphy in the Netherlands based on analyzed cores from areas within and adjacent to the Rhine Delta. This area has been subsiding during the entire late Cenozoic, resulting in a continuous, or almost continuous, sediment stratigraphic record. (Modified from W.H. Zagwijn, 1975.) Blue: cold, glacial or stadial periods. Red: warm, interglacial or interstadial periods. A revised graph has been published by W.H. Zagwijn (1985). He introduces a Bavelian Stage between the Menapian and the Cromerian.

sections of the four glaciations may indeed represent glaciations of ages different from the four originally indicated in Fig. 2–1. Therefore, there is a tendency today to abandon the names of at least the oldest, more-problematic glaciations and use the marine isotope-stage numbers shown in Fig. 1–19 (and Fig. 2–2) instead.

Old glacial deposits on land have been observed as beds exposed in stratigraphic sections, for example, in sea bluffs, river cuts, and drill cores. However, none, or very little, of the original topography of the old drift surfaces is generally preserved. Hills and ridges composed of morainic and glaciofluvial deposits were certainly formed during the earlier glaciations, but they were probably levelled over time by gravity and partly removed by subsequent glacial and glaciofluvial erosion together with erosion by other processes during the following glacial and interglacial periods. The glacial deposits from the last glaciation (Wisconsin/Weichselian) usually have a fresh, relatively sharp surface topography, still displaying steep-sided hills and ridges. The next-oldest drift sheets, from the Illinoian/Saalian Glaciation, generally have a much more subdued surface topography. In some areas where the moraine ridges contain much fine-grained material, as, for instance, in the Midwest area

Fig. 2–3. The maximum extent of the North European Ice Sheet during various late Cenozoic glaciations. 1: The Weichselian Glaciation. 2: The Warthe phase of the Saalian Glaciation. 3: The Saalian Glaciation. 4: The Elsterian Glaciation.

B: Berlin. W: Warsaw. M: Moscow. P: Prague. X: Erratics at Belchatow (see Fig. 2–6). XX: Erratics northwest of Celle (see Fig. 1–9). Y: Erratics at Horni Rasnice in the Czech Republic (see Fig. 2–6). The orange area was never glaciated.

Fig. 2–4. Rhombporphyry erratics of the characteristic rhombporphyry bedrock which outcrops in the Oslo region – the only region in Europe where this kind of bedrock exists.

A: Erratic at Stensigmose in Denmark lies at the foot of a sea cliff where glacial deposits from several glaciations are exposed. X: a camera.

B: A small rhombporphyry erratic at Emmerlev in southern Denmark. It lay at the foot of a sea cliff with Saalian till.

C: Erratic from glacial deposits north of Hamburg (from a rock collection at Geolgishes Landesamt Schleswig-Holstein). Rhombpophyry erratics are excellent indicator erratics. (See distribution area on Fig. 2–6E.)

during at least two earlier glaciations. During its maximum extension, the North European Ice Sheet covered all of Scandinavia, parts of Holland, much of Germany, most of Poland, the Baltic countries, and much of Russia including Moscow (see Figs. 2–3, 2–6). In addition, it is believed that the North Atlantic Ocean was partly covered with ice shelves or sea ice (pack ice) all the way south to the coasts of Spain, Portugal, and New England in USA, at least during some, if not all, of the main glaciations.

North America was glacier covered over its northern two-thirds by two essentially contemporaneous but separate ice sheets, the Cordilleran and Laurentide. The smaller Cordilleran Ice Sheet covered the western mountain districts and the larger Laurentide the remainder, and at times of maximum glaciation they merged. Some of the pre-Wisconsin-age Laurentide ice sheets were somewhat more extensive than the Wisconsin-age Laurentide Ice Sheet in the Midwest region, but probably no larger in other parts (Fig. 2–5), while the earlier Cordilleran ice sheets appear to have been smaller than that of Wisconsin age.

of the United States, the moraines of those ages have been almost completely levelled. However, in other areas where the morainic material is coarse grained, as, for instance, in parts of the Alps and the Rocky Mountains, even the Illinoian/Saalian moraines still retain good topographic expression, although they are generally somewhat subdued. Even the Saalian Warthe moraines on the plains in Poland and northern Germany (Fig. 2–3) are relatively well defined, yet they consist of considerably subdued ridges.

The extent of the old ice sheets

In North America and northern Europe the extent of the ice sheets was generally larger than that of the Weichselian/Wisconsin expansion

Fig. 2–5A. The maximum southern extent of the Cordilleran and Laurentide ice sheets during the Wisconsin and pre-Wisconsin glaciations. Recessional marginal positions of the Late Wisconsin Ice Sheet are added: 1 Ka = 1000 years. (Modified from R.F. Flint, 1971.)

Fig. 2–5B. Dispersal fans of boulders derived from unique bedrock sources (red) in eastern North America. The fans demonstrate glacier flow from the northwest over the New England area. (Modified from R.F. Flint, 1971.) Similar fans have been observed in many parts of the formerly glaciated North America.

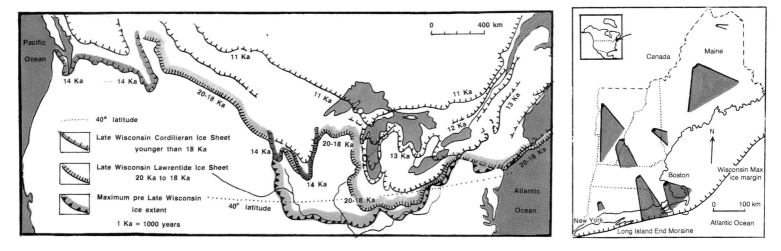

Holocene soil
Weichselian cover sand

Saalian till with large
Fennoscandian erratics,
brown oxidized in upper
part (Eemian oxidation)

Saalian glaciofluvial
deposits

Fig. 2–6A-D. Fennoscandian erratics in the Bel-chatow Pit, southern Poland. A: Sediment beds exposed in the upper part of the pit. The erratics shown on the other pictures lay in the Saalian till, and all large erratics are of Fennoscandian origin; several are glacially striated (B), and the pink Rapakivi granite (C) is from Aaland. Altogether, eight till beds, supposedly of Saalian and Elsterian age, have been observed in the Belchatow Pit.

Fig. 2–6E. Characteristic indicator erratics from Fennoscandia are found within the entire area that was covered by the North European Ice Sheet. The source areas for four of the best-known types of erratics are shown on the map together with their fan-shaped distribution areas. 1: Location of the Belchatow Pit (see pictures A-D). 2: Location of the erratics shown on Fig. 1–9. 3: Locations of the rhombporphyry erratic shown on Fig. 2–4. 4: Location of erratics shown on Fig. 2–6F.

Fig. 2–6F. Fennoscandian erratics in a pit near Horni Rasnice in the Czech Republic, close to the Polish border (see Fig. 2–6E). Most of the large erratics are of Fenno-scandian origin. Exceptions are some fairly large chert boulders (marked with x), which probably originate from Mesozoic beds at or near the south coast of the Baltic Sea.

Y: Saalian till (A) which overlies glacitectonized beds with folds indicating a transport (shear stress) in a southerly direction (not shown on the picture).

B: Sand beds, mostly covered with slumped sand, or erratics from A.

C: Bouldery, poorly sorted gravel beds which represent ice-contact outwash beds. Note the shape of many boulders. They are subangular to subrounded, which is typical for glacially transported erratics. This shows that they were transported by glaciers most of the long way from Fennoscandia, and only a very short distance by glacial rivers.

Z: Erratics derived from A and C. A faint glacial striation was still preserved on the large, dark gabbroic erratic.

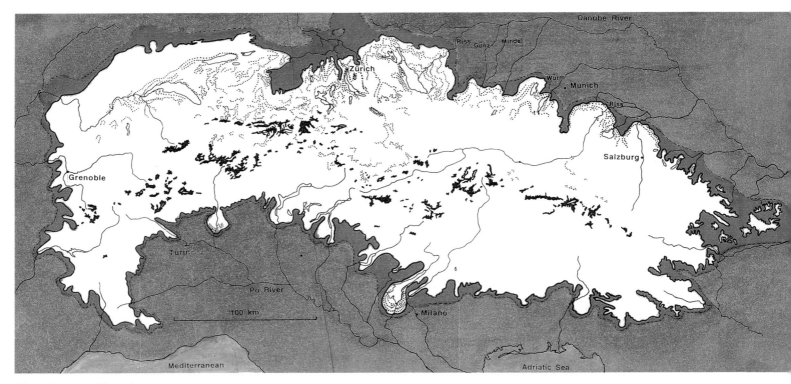

Fig. 2–7A. Glaciations in the Alps. Dashed line: The maximum extent of the pre-Würmian glaciers, roughly outlined. Blue: area covered by pre-Würmian glaciers. Heavy line: The maximum extent of the Late Würmian glaciers, about 20 000–18 000 years ago.

Dotted lines: Maginal moraines representing younger Late Würmian glacial substages. Only a few of the known moraines are plotted. Black: existing glaciers. Orange: area which was never ice covered. (Modified from various sources.)

The glacial history of the Alps

Glaciation of the Alps was a highly important factor in determining the total environment of ice age Europe, and the glacial ice age theory originated in the Alps. Here the subdivision of the Quaternary into four glacials and three interglacials, proposed in 1877 by A. Geikie for Britain, was further developed and applied. The subdivision in the Alps was introduced by A. Penck and E. Brückner in their famous publication of 1901–9, "Die Alpen im Eiszeitalter". They did much of their field work on glacial and glaciofluvial deposits in valleys on the north slope of the Alps, and they named the four glaciations after four rivers which radiate out from the central high mountains of the

Alps and join the Danube River. Figure 2–7B illustrates the suggested relationship between the four glacials and outwash terraces and outwash bodies in those river valleys. It was later suggested that glacial deposits of at least one older glaciation, the Donau Glaciation, exist in the Alps.

Since the pre-Würmian (pre-Weichselian) morainal chronology of the Alps is based primarily upon morphology and observed differences in the weathering of deposits, and to a lesser extent on their stratigraphy and absolute ages, many scientists seriously question the ages assigned to some of the older moraines in particular. The central Alps were not completely ice covered during the glaciations. Numerous peaks and ridges rose above the ice surface as nunataks, while larger ice-free areas lay between the outlet glaciers in the main valleys. Fig. 2–7A presents the main outline of glacial

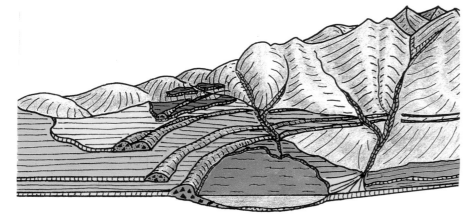

Fig. 2–7B. Evidence of four glaciations in the Alps. A generalized sketch indicating the evidence presented by Penck and Brückner (1901–11) from valleys on the north flank of the Alps (see Fig. 2–7A). Brown ridges: marginal moraines. Orange: outwash terraces of Würmian age. Yellow: outwash terrace of Rissian age. Green: outwash terrace of Mindelian age. Red: outwash of Günzian age. Brown: post-Würmian terraces.

phases during the last glaciation, the Würmian. During some of the earlier glaciations the ice cover was more extensive, as indicated.

The climate, fauna and flora of the old glacials

What do we know about the climate, fauna and flora of the older glaciations? We know that the climate was similar to that of the last glaciation, although it could have been slightly more severe during the more extensive glaciations. At that time the eustatic global sea level dropped nearly 150 m in response to global glacier accumulation, causing enormous expansions of the land areas. Important land bridges then connected North America with Asia, Borneo with Australia, and Europe with Africa. In turn, these allowed extensive migration of both plants and animals, including early humans.

In addition, several groups of animals that lived on the plains and in the forests of Europe and North America experienced a striking evolution from the warmth-adapted late Tertiary species to cold-adapted late Quaternary species. The evolution of the wooly elephant (*mammoth*) and wooly rhinoceros are good examples of these changes. Along with evolutionary changes there were also extinctions. Most of the mammalian species that lived 2.5 million years ago became extinct during the late Pliocene and the Pleistocene, for example, the famous European Villafranchian fauna and the North American Blancan fauna, which lived 2–3 million years ago. The flora changed as well, and many late Pliocene/early Pleistocene species do not exist today. However, the vegetation pattern and species of the earlier glaciations were most likely quite similar to those of the last glaciation.

Even though we know a great deal about the conditions during the earlier glacials, the details from the older phases are still fragmentary, and absolute dates are much fewer and less accurate than those for the last glaciation, which began after 115 000 years ago. However, there are also great similarities in glacial and climatic conditions and in the physical-geographic land conditions during most of the glaciations. Therefore, we will focus on more detailed description while presenting the model for the last glaciation.

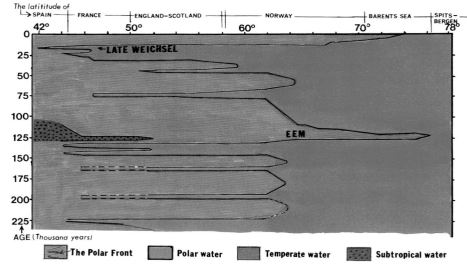

Fig. 2–8A. Latitudinal fluctuations of the Oceanic Polar Front in the North Atlantic during the last 250 000 years. (Modified from T.N. Kellogg, 1976.). The location of the Oceanic Polar Front corresponds with the southern limit (2°C) of cold Arctic surface water (see p. 182). The graph is based on observation of planktonic micro-fossils in many cores collected from the ocean floor. The fossils render information about the surface-water conditions. Note how well this graph corresponds with the presented oxygen-isotope graphs, and how far to the north the Polar Front was located during the Eemian/Sangamonian Interglacial. The studies indicate that the E/S Interglacial probably had the warmest surface water of all recorded late Quaternary interglacials.

Interglacial phases (warm phases) older than 130 000 years

Interglacials, in contrast to glacials, are characterized by warm climates and dense vegetation over much of Europe, North America, and several other parts of the world. There the vegetation cover prevented much soil erosion, and consequently the rivers and the wind transported little material in general. Therefore, relatively little clastic (inorganic) sediment was generally deposited either on land or in the sea during the interglacials. What we do find on land are primarily organic lake deposits and buried soils (paleosols) which represent the weathering of former exposed interglacial land surfaces. In addition, we find high-lying beaches and marine terraces along many coasts, since the sea level was higher than today due to additional melting of glacier ice. The old interglacial beaches lay no more than 20 m higher than the present, but in tectonically unstable regions, such as Italy and California, interglacial beach deposits have been found several hundred metres higher. In general, the best-preserved and best-studied features represent the warmest phase of the last inter-

Fig. 2–8B. Temperature fluctuations, departures from present-day temperature, at Vostok Station on the South Pole Plateau (see Fig. 1–24). The large and rapid climate changes at the start and end of the Eemian Interglacial are well displayed on this graph, and they are generally very clearly recorded on most other climate graphs which cover the Eemian. (Modified from J. Jouzel and others, 1987.) Blue: cold glacial phases. Red: warm, or relatively warm, phases.

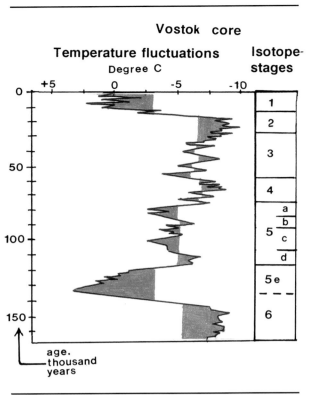

The warm Eemian/Sangamon (E/S) Interglacial about 130 000–115 000 years ago

The E/S Interglacial is usually correlated with deep-sea isotope stage 5e (see Figs. 2–8B, 2–17). However, there has been considerable discussion about where to draw the boundary between the Eemian and the Wisconsin/Weichselian Glaciation, and some scientists have considered that the entire isotope stage 5, from 130 000 to 75 000 years ago, represents the E/S Interglacial. This is a matter of definition, and the concept that E/S is equal to isotope stage 5e seems to fit best the general definition (see Glossary); thus, it will be used in this book.

Our knowledge about the E/S Interglacial has improved vastly during recent decades. New dates and modern studies of cores from both land areas and sea floors have shown that:

1. The glacial retreat and climatic change which led from the previous glacial to the E/S Interglacial was rapid.
2. The warm E/S Interglacial (5e) lasted a maximum of 20 000–15 000 years.
3. A warm peak with mid-latitude mean summer temperatures about 2°C above the present, and with glaciers smaller than at present, occurred in the early part of the E/S about 125 000 years ago, a situation which lasted only a few thousand years.
4. A gradual cooling followed the warmest phase of the E/S.
5. The Eemian climate in western Europe was moist.
6. The final climatic cooling and glacial expansion which led from the E/S Interglacial to the Weichselian/Wisconsin Glacial was rapid.
7. The pattern for the E/S shown on the oxygen-isotope graphs corresponds well with the above-mentioned description, and it is very similar to some of the patterns for earlier interglacials, suggesting a similar climatic and glacial evolution.
8. The pattern for the E/S changes of vegetation and shifts of vegetation zones shown in pollen diagrams also corresponds with the previously mentioned descriptions, and with changes recorded in the oxygen-

glacial, the mid-point of which was 125 000 years ago. They will be selectively discussed below.

Fossiliferous pre-*Eemian/Sangamon* interglacial deposits, older than 130 000 years, have been recorded at many localities, in both marine and terrestrial environments. The marine beds from the penultimate interglacial, the *Holsteinian/Yarmouth* Interglacial, reveal that the sea temperatures were similar to or slightly cooler than those during the last interglacial, the Eemian/Sangamon (see Fig. 2–8A).

The flora of the Holsteinian/Yarmouth Interglacial in North America and northern Europe is recorded in deposits both from lakes and bogs in areas near the Weichsel/Wisconsin maximum ice limit, and in areas further away from this limit. The pollen content in cores from many lakes and bogs gives a continuous record of the changes in vegetation as the vegetation zones passed across the particular bog/lake area. Figure 2–12 shows a pollen diagram through the Holsteinian time in northern Germany. The striking similarities and differences between the vegetation development shown in the Holsteinian diagrams and the vegetation development of the Eemian Interglacial will be discussed in the next chapter.

isotope graphs. In addition, these changes are in general similar to the vegetation-change patterns during older interglacials, although some characteristic differences can be observed. The Eemian patterns in Europe are similar to the vegetation patterns for the Holocene, from 10 000 years B.P. to the present.

9. The E/S mammalian fauna was, in part, the same as the fauna of the Holsteinian/Yarmouthian Interglacial. However, a considerable element of the Eemian fauna became extinct and is missing from the Holocene and present-day fauna. The E/S global sea level was in the order of 4–6 m higher than at present.

10. The Eemian soils are frequently very spectacular (Figs. 2–9, 2–10).

The vegetation of the E/S Interglacial, 130 000–115 000 years ago

Europe. Eemian Interglacial vegetation. The Eemian regional vegetation zones shown in Fig. 2–11 represent the warmest (125 000–120 000 years B.P.) phase of the interglacial. Note that western Europe was forested all the way north to the Arctic Ocean, and that the regional forest

Fig. 2–9. Eemian Interglacial peat (6) and Early Weichselian, Brörup Interstadial, peat (8) interbedded with what may be fluvial sand at Schalkolz in northern Germany. (Photo by Jan Mangerud.)

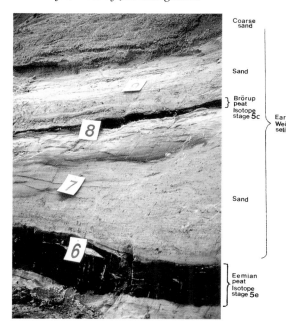

Fig. 2–10. Eemian Interglacial soils can be very spectacular. They are the result of weathering and other processes connected with the vegetation. Buried Eemian soils are frequently very striking in areas beyond the regions covered by the W/W glaciers.

I: Eemian (to early Weichselian) buried chernozem soil in the loess area in southern Poland. A: Holocene soil. B: Weichselian loess. C: Chernozem soil with a black organic horizon above a brown-oxidized zone. The organic wedges and pockets which project down into the Saalian loess could partly have been formed in connection with root systems. D: Saalian loess. E: Other Saalian sediments. Note the thick unit of Weichselian loess.

II: Eemian podsol soil on the west coast of Jutland, Denmark. The soil is formed on Saalian till and is overlain by Weichselian sand (S). A_1: The A_1 horizon with accumulated black organic material. A_2: The A_2 horizon where all dark minerals have been leached (dissolved) and the ions have been transported downwards with the percolating water. B: The B-horizon where dissolved matter from A_2 was deposited. C: A thin, dark-brown manganese zone. D: A wide, brown zone with precipitated iron oxide.

Note the many white wedges and pockets where the leaching has penetrated deeper, and the well-developed ice-wedge cast below X. The corresponding ice wedge was formed in a cold Weichselian period. In both Fig. I and Fig. II the soil resembles so-called "Taschenböden" (pocket soils) where the formation of the pockets has been ascribed by some scientists to frost action.

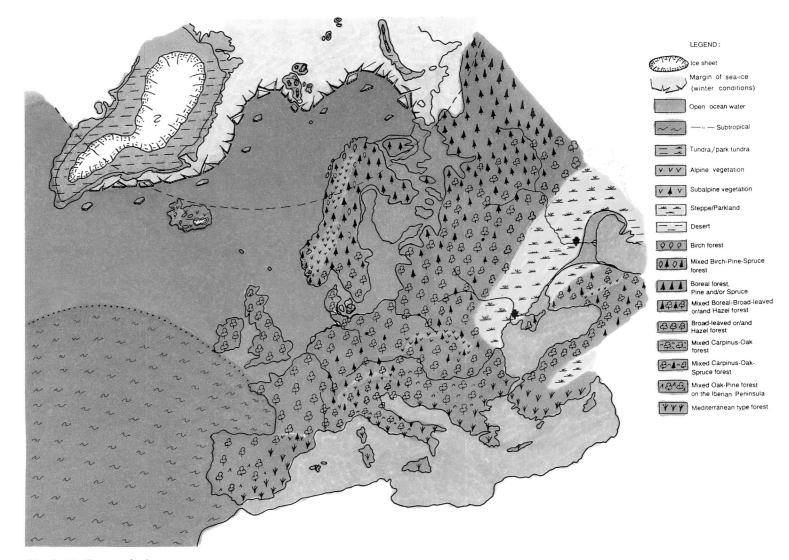

LEGEND:

Ice sheet

Margin of sea-ice (winter conditions)

Open ocean water

— " — Subtropical

Tundra/park tundra

Alpine vegetation

Subalpine vegetation

Steppe/Parkland

Desert

Birch forest

Mixed Birch-Pine-Spruce forest

Boreal forest, Pine and/or Spruce

Mixed Boreal-Broad-leaved or/and Hazel forest

Broad-leaved or/and Hazel forest

Mixed Carpinus-Oak forest

Mixed Carpinus-Oak-Spruce forest

Mixed Oak-Pine forest on the Iberian Peninsula

Mediterranean type forest

Fig. 2–11. Europe during the Eemian Interglacial, about 125 000 years ago. The temperature over much of Europe was about 2°C warmer than today, the forest covered even the northernmost parts of the continent, and subtropical water migrated northwards to a position west of Ireland. Observe that Fennoscandia was a large island! The size of the Greenland Ice Sheet at that time is highly questionable, and it could even have been larger, or considerably smaller, than indicated.

zone boundaries lay slightly to the north of the present and Holocene boundaries. The boundary between warm, subtropical water and temperate water in the North Atlantic lay considerably north of the present-day boundary (see Figs. 2–11, 2–69) .

Figure 2–12 presents a generalized Eemian pollen diagram for southern Scandinavia, northern Germany and Holland. The diagram shows the changes in vegetation through the Eemian Interglacial, and it is representative for a large part of northern Europe, present northern Germany, Holland and northern Poland. An early warm phase with broad-leaf/mixed oak forest dominated by oak, and a very distinct hazel maximum, has been recorded from all over northern Europe up to southern Scandinavia. A long period of gradual cooling took place

during the last half of the Eemian, in which the broad-leaf forest was gradually replaced by a Boreal spruce-pine forest in most of northern Europe. In this late period the amounts of spruce pollen reached very high values in most areas, except in Britain. For comparison a pollen diagram for the Holsteinian Interglacial is also presented in Fig. 2–12. This diagram represents the same area as the Eemian diagram, and it shows a considerable number of similarities with the Eemian diagram, in addition to some important differences. For instance, the distinct early oak/hazel maxima and the late spruce maximum are missing in the Holsteinian diagram. The broad-leaf/mixed oak forest-pollen values are relatively low and those for spruce are low throughout most of the Holsteinian.

Pollen diagrams which present all observed tree taxa and all herbs and shrubs can be used

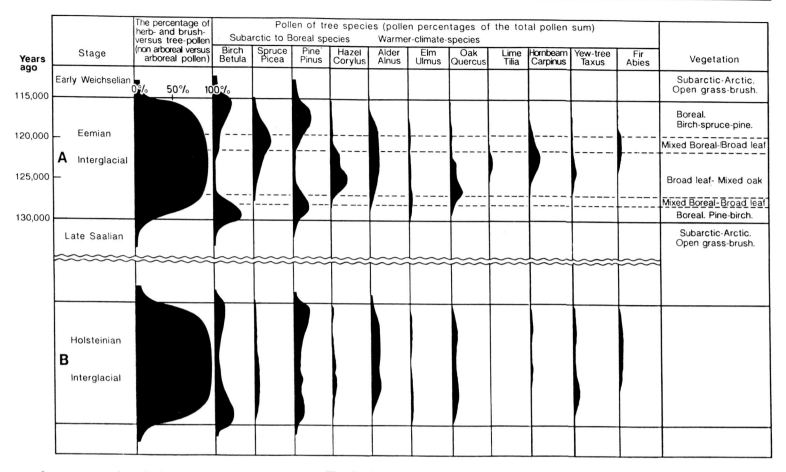

to draw more detailed conclusions about the climate. For instance, the presence in southern Denmark and northern Germany of *Ilex, Hedera, Viscum* and *Buxus* during the warm phase of Eemian and of *Ilex, Taxus* and *Buxus* during warm phases of Holsteinian indicate that the climate there was warm and moist during the two phases.

North America Sangamon Interglacial vegetation. The relatively large North American continent comprises many very different vegetation zones, and it is not possible to present a single Sangamon pollen diagram which is representative for the whole. In fact, very few diagrams are available for the Sangamon Interglacial in North America. The few diagrams available from the southern part of the United States show relatively small changes in vegetation through the interglacial. From more northern regions, there are a few diagrams from the West Coast and the Midwest, but none from the eastern part of the continent. The few existing ones from the Midwest show a very brief Subarctic-Boreal, spruce-pine zone at the

Fig. 2–12. A: A generalized pollen diagram recording a typical Eemian forest-tree succession in the region comprising northern Germany, northern Poland and southern Denmark. B: A corresponding diagram for the Holsteinian Interglacial. Black: tree pollen.
The same tree species are present in both interglacials, but the vegetation patterns are characteristically different.
More complete pollen diagrams which include records of many plant species give more detailed information about the vegetation and climate. (These diagrams are based on several diagrams presented by different scientists, including S.T. Andersen, 1965; B. Menke and R. Tynni, 1984; and B. Menke, personal communication.)

beginning and the end of an interglacial which appears to have been somewhat warmer than today. The mountain regions in the west show longer transition periods, characterized by open Arctic-Subarctic vegetation before and after the Boreal vegetation period which dominated the central part of the Sangamon.

Animal life of the E/S Interglacial, 130 000–115 000 years ago

Europe. Mammalian fauna was in several ways conspicuously different from the present fauna. Animals such as the "forest elephant",

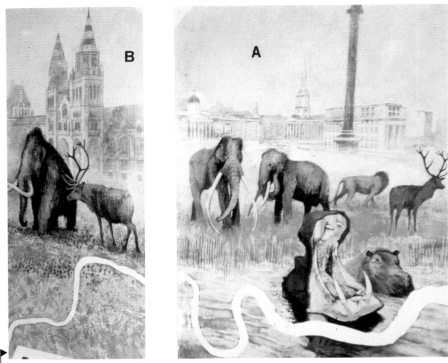

Thames River

Fig. 2–13. A: Trafalgar Square in London, 125 000 years ago. Fossils of hippopotamus, forest elephant, lion etc. representing a typical Eemian fauna were found in the sediments of Thames River terrace at Trafalgar Square.

B: London, 20 000 years ago. Fossils of an Arctic fauna with mammoth, reindeer etc. lay in Thames River terrace sediments of Devensian (Weichselian) age. (From an exhibition at the Natural History Museum, London.)

the "forest rhinoceros" and hippopotamus lived in parts of Europe, including England, Germany and France. The reconstruction in Fig. 2–13 is based on fossils found in London, and it illustrates the mega-fauna there during the E/S.

Besides the exotic, now-extinct mammals, many of the present-day mammalian species, such as elk, moose and deer, lived in the broadleaf forest which covered much of Europe. An interesting cold-climate fauna, which was typical for much of Europe during the glaciations, lived in the Arctic and Subarctic regions of northernmost Europe during the warmest phase of the E/S. (See next section for more details.)

North America. The Sangamon mammalian fauna in North America was also different from the present-day fauna. Many species have become extinct. This is particularly true for many of the large mammals, such as the *Columbian mammoth*, the *Ground sloth* and the *Giant bison*, which were common in the Midwest. It

is interesting to notice also the difference between the North American and the European E/S fauna. The Bering Land Bridge was exposed during only the glacials, and therefore no warm-climate species could cross this bridge during interglacials. Hence, the warm-climate faunas on the two continents developed independently during late Quaternary time.

Other faunas used in reconstructing the E/S climate. Fossils of several other animal groups are also very important for reaching conclusions about the Eemian climate. For instance, in marine sediments, fossils of both molluscs and corals, and of micro-organisms such as forams, diatoms and radiolaria, record warm-water conditions during the E/S. In terrestrial sediments, fossil beetles also record a warm E/S climate.

The E/S marine zones

The regional sea-surface temperature zones migrated in response to the climatic changes in about the same way as the vegetational zones on the continents. The subtropical warm-water zone was pushed northwards in the North Atlantic during the E/S Interglacial and had a northern limit to the west of Scotland. The zone of polar water, which is generally ice covered during the winter, was pushed northwards also, to a position closer to Spitsbergen than it is today (see Figs. 2–8A, 2–11). Some scientists have called the southern limit of this zone the "Polar Front", or the "Oceanic Polar Front", and it is marked by about 2°C sea-surface temperatures. Regional changes in *planktonic* marine fauna and flora recorded in deep-sea cores, together with regional changes in the marine fauna recorded in coastal waters, are generally used to distinguish the different sea-surface temperature zones (see Fig. 2–8A).

The E/S high marine sea level

The E/S eustatic sea level was 4–6 m higher than today, according to most evidence. Shorelines, terraces, and other shore sediments containing a warm-water Eemian fauna have been found at about that level on the stable coasts of several continents outside the glaciated regions, and away from the tectonically unstable regions. At that time many present-day, low coastal plains around the continents were flooded (Fig. 2–14), and many of the world's larger cities would have been drowned if they had existed during the E/S Interglacial.

What caused the E/S rise in sea level? The warming of the world's ocean water accounted for a slight expansion of the water and a related slight rise in sea level. However, the large rise in sea level of 4–6 m above the present resulted mainly from the melting of glaciers on the continents. Therefore, scientists have speculated about which glaciers melted and caused the rise in sea level. There are two possible candidates for this change, the Antarctic Ice Sheet and the Greenland Ice Sheet.

Cores collected to date the basal part of the Greenland Ice Sheet suggested that most of the ice is of Weichselian age, and this together with theoretical glaciological calculations have led many scientists to believe that much of the Greenland Ice Cap vanished during the E/S Interglacial. However, recent analysis of a core to the base of the ice sheet in central Greenland indicated that some ice was there during the Eemian, and the amount of ice melted was much too small to account for the relatively large Eemian rise in sea level.

Theoretical estimates and calculations have been made in favor of a disappearance of parts of the Antarctic Ice Sheet also, in particular the marine-based West-Antarctic Ice Sheet. Probably a combined reduction of the Greenland and the Antarctic ice sheets caused the E/S rise in sea level.

Can the sea level rise again?

The Eemian climate was warmer than today, and this raises the serious question whether an increased warming in the near future could cause a similar rise in sea level. Considering the catastrophic results of such a rise, the question has been much in focus lately in the discussion of the possible results of global warming, presumably due to the increase of "greenhouse gases" in the atmosphere.

The climate was slightly warmer than today during the warmest Holocene time, 8000–4000 years ago, but it did not result in a very high sea level. In addition, the pattern of the climatic

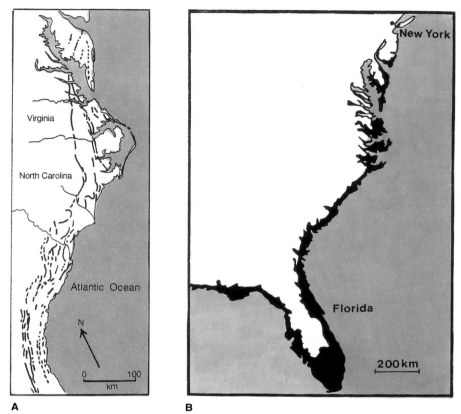

A   **B**

graph for the latest 200 000 years suggests a trend towards a cooling in the near future rather than a warming of the climate. However, the increase of carbon dioxide (CO_2) and other "greenhouse gases" in the atmosphere, primarily caused by human activity, could be the cause for an observed small warming (greenhouse effect), and if this increase continues there is the possibility that the expected future climatic pattern will be different. Some scientists have postulated that the expected cooling at the transition to the oncoming glaciation will be replaced by a super-warm period of short duration, and thus the first cooling phase of the next glaciation will be delayed. In that case the warming may result in the severe rise of sea level feared by many scientists, but this scenario is highly speculative.

The shores and prehistoric human cultures

Early humans frequently lived on or near the shores, and artifacts in beds from different human cultures are often closely related to the shore deposits, such as those of the E/S Interglacial (see p. 97).

Fig. 2–14A. Interglacial beach ridges and other shore features (heavy lines) on the east coast of USA. The youngest, lowest-lying features, including the Surrey and Suffolk scarps, represent the last interglacial (Sangamon-Eem). (Modified from R.F. Flint, 1971.)

Fig. 2–14B. Sea-covered coastal districts (black) during the Sangamon-Eem interglacial, about 125 000 years ago. The sea level was 4–6 m higher than today. During the Wisconsin glacial maximum, about 20 000 years ago, the sea level was 100–120 m lower than today along this coast. (Modified from P. Mac-Clintock and H.G. Richards, 1936.)

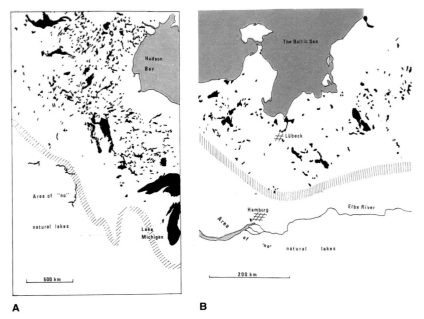

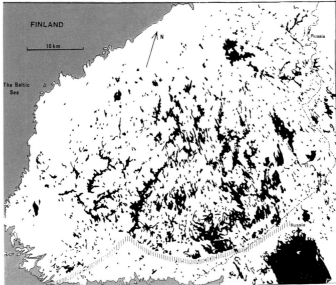

A   **B**   **C**

*Fig. 2–15. Lake landscapes are characteristic of areas that were covered by the gla-
ciers/ice sheets of the W/W glaciation. The lakes (black) lie in glacially eroded rock
basins or in basins dammed by glacial deposits. A: Lakes in a part of North America.
B: Lakes in a part of northern Germany. C: Lakes in a part of Finland.*
*Numerous small lakes are too small to show on these map scales, and note that
practically no natural lakes exist in areas beyond the Wisconsin/Weichselian ice limits
(shaded belts on A and B). The shaded belt on C is the Salpausselkä I end moraine of
Younger Dryas age.*

The Weichselian/Wisconsin (W/W) Glaciation, 115 000–10 000 years ago

General review

The last great ice sheets had their largest extension in much of North America and Europe
21 000–17 000 years ago. At that time Long
Island, New York, and the hilly terrain adjacent
to Berlin in Europe, were formed as end
moraines along glacier margins. However, the
last glaciation started with a global cooling
about 115 000 years ago, according to the most
accepted data, and the last remnants of the
large North American Ice Sheet finally disappeared as late as about 8000 years ago, and
those of the North European Ice Sheet about
8500 years ago. The deposits from the last
glaciation are generally strikingly "fresher"
than deposits from the older glaciations. The
end moraines are often topographically well
expressed as steep-sided hills and ridges, and
most of the numerous glacially formed lake
basins of the world lie within the area covered
by the glaciers of the last glaciation (see Figs.
2–15, 2–16). Therefore the glacial drift of the
last glaciation has been informally called the
"new drift" and the drift of the older glaciations the "old drift" in many areas.

In fact, the presence of numerous lakes is
one of the most striking features of the areas
covered by the W/W ice sheets. The lakes
generally lie in depressions in the W/W drift
in the marginal zones (Fig. 2–16), and in glacially eroded rock basins in the more centrally
located (proximal) zones. In areas beyond the
outer margin of W/W ice sheets, almost all of
the formerly existing lake basins in the old
drift sheets are filled and covered with younger deposits, frequently with distal W/W deposits (see Fig. 2–15).

The changing climate of the W/W Glaciation

The intensive stratigraphic studies which have
been carried out recently have shown that the
climate and glacier fluctuations during the last
glacial have been more varied and complex
than previously believed. Both the climate and
the glaciers have fluctuated many times during
this glaciation. Significant fluctuations of this
kind within a glacial are called stadials and
interstadials for the cold glacial phases and the
warm phases. The climate of an interstadial is
usually considerably cooler than that of an

interglacial. However, for a short period it can be about as warm as the interglacial climate. If that period had been long enough for the world's glaciers to shrink to about present-day sizes and for the vegetation zones to reach approximately present-day positions, it would have been called an interglacial. (See discussion in the Glossary for details.)

The last glaciation has been frequently divided into an early part about 115 000–75 000 years ago, a middle part about 75 000–25 000 years ago, and a late part about 25 000–10 000 years ago. However, there never has been a generally accepted definition of the three parts, and it has now become very common to refer to the more precisely defined deep-sea zones in the discussions concerning the age of the terrestrial deposits. Note that the time scale for the older deep-sea zones is based mainly on Uranium-series dates, and they are frequently older than the radiocarbon dates presented earlier for the same zones.

Information about climatic fluctuations comes from studies of stratigraphic sections, most of which are cores from lakes and bogs or from the large ice sheets and the ocean floor. Figures 2–17, 2–18 and 2–8A show some general graphs from various parts of North America, Europe, and the North Atlantic Ocean during the last glaciation. The best sections and graphs are in general from areas that lay outside the glaciated regions, but important information does come from the glaciated regions also. In most of the records presented the inferred climatic trends are very similar, which suggests that the climate changed in much the same manner in Europe and North America. Even in other parts of the world the major climatic trends seem to correspond fairly well with the trends in North America and Europe. In general the intensity of atmospheric circulation increased during the glacials and stadials in most glacial and periglacial regions.

Europe, 115 000–25 000 years ago

The terrestrial record for the 90 000 year early and middle part of the Weichselian Glaciation is based mainly on information from the analysis of stratigraphic sections in various parts of Europe. The results from the studies in

Fig. 2–16. Hummocky lake landscapes are typical for the glaciated regions.

A: Late Wisconsin hummocky morainic terrain with many small moraine-dammed lakes at Coteau des Praires in South Dakota. (Photo by John Shelton.) This kind of terrain is rather typical for many areas which were covered with Late Wisconsin/Weichselian ice sheets, both in North America and in Europe. However, they are usually forested and therefore not as well displayed as on this picture.

B: Lakes in rock basins formed by glacial erosion on Røyevidda plateau in Finnmark, northern Norway. The rock basins are eroded in hard, crystalline bedrock. (Photo by Tor Schulstad.) Lake basins on flat terrain like this were eroded by ice sheets, and they are shallow. Considerable parts of both the Canadian and the Fennoscandian bedrock shields have similar terrains with numerous lakes.

general give a rather consistent picture of the climate fluctuations, but the information about the corresponding glacier fluctuations is still rather incomplete. Glacial deposits have been recorded in Scandinavia from all of, or most of, the cold stadials, but the extent of the glaciers during the stadials and interstadials is still controversial. Figure 2–17 presents a rather

Fig. 2–17. Climate fluctuations recorded in three areas of western Europe, from south (graph A) to north (graph C) combined with a deep-sea oxygen-isotope graph (modified from B. Andersen and J. Mangerud, 1990).

Red: warm, or relatively warm, climate. Blue: cold, or relatively cold, climate.

A: A graph for northern France based on vegetation changes recorded in a pollen-analyzed bog section at Grand Pile (modified from G.M. Woillard, 1978).

B: A graph for Holland-northern Germany based on several pollen-analyzed sections (modified from K.E. Behre, 1989).

C: Climate and glacier fluctuations in Fennoscandia, based on studies of both sediments and fossils in many stratigraphic sections within the glaciated areas, where each section generally covers a relatively short time span. The correlation between the sections is frequently problematic, and therefore this graph is not as well founded as the others.

D: A deep-sea oxygen-isotope graph (modified from Marthinson and others, 1987).

The graphs A, B, and C show the trends rather than the exact amplitudes of the climate fluctuations, and the ages of the fluctuations older than 50 000 years are based mainly on correlation rather than absolute dating.

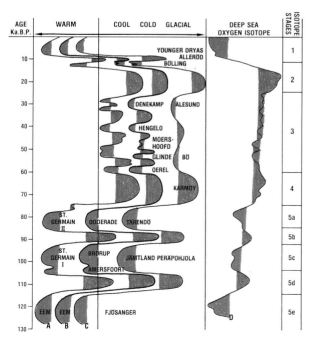

speculative reconstruction of these ice-sheet fluctuations.

Instead of presenting a review of the many available climate records, we will focus primarily on the Grand Pile record from northeastern France, which is one of the most complete land records available. The youngest parts are rather well dated by radiocarbon methods, and the ages of the older parts are based on correlation with the oxygen-isotope stages in the deep-sea record. The graph (A) presented in Fig. 2–17 is based on pollen analysis of a bog section, and it shows fluctuations between times with dense forest vegetation of various types and times with open tundra or prairie vegetation. These vegetation changes reflect the climate changes. For comparison a graph (B) based on pollen analysis of sections in northern Germany is added together with a similar graph (C) for Fennoscandia. However, the latter graph is based on very fragmentary and incomplete information.

The early Weichselian (isotope stages 5a–5d), 115 000–75 000 years ago

The Grand Pile section shows three very warm interstadials and two cool stadials during the early Weichselian (isotope stages 5a–5d). Because of the very warm character of the interstadials, some scientists want to include the entire isotope stage 5 in the Eemian Interglacial. However, the corresponding interstadials and stadials in northern Europe are considerably colder, and there is evidence of good-sized glaciers in Scandinavia during the stadials. The two main interstadials of Grand Pile, the St. Germain I and II, have been correlated with, respectively, the Amersfoort-Brörup and the Odderade interstadials in northern Europe (Fig. 2–17).

The middle Weichselian (isotope stages 4 and 3), 75 000–25 000 years ago

Isotope stage 4 (75 000–60 000 years ago). The climate cooling which followed isotope stage 5a (Odderade) was rapid, and it marked the start of a very long cold period. During the major cold peak of isotope stage 4, the tundra/prairie covered central and northern Europe, and the glaciers expanded to cover much of Scandinavia. The Oceanic Polar Front moved south to the latitude of southern France and extended westwards towards North America. To the north of this latitude most of the North Atlantic Ocean was covered with pack ice during winters.

Isotope stage 3 (60 000–25 000 years ago). The climate remained cold during most of isotope

Fig. 2–18. Glacier fluctuations in North America. Blue: cold glacial periods. The glacier fluctuations in North America during the Wisconsin Glaciation, based on various sources, including A. Dreimanis and P. Karrow (1972). The stadials and interstadials in the lower part of the diagram are not well dated. They are given ages by correlation with well-dated isotope stages. The St. Pierre Interstadial and Nicolet Stadial are correlated with isotope stages 5a and 5b. However, there are no well-defined interstadial and stadial events, so far recorded, which correspond with isotope stages 5c and 5d. In most reconstructions (a) these two events are included in a late part of the Sangamonian. In reconstruction (b) a small 5d glacier advance is indicated, but this is highly speculative. In addition, it is possible that St. Pierre and Nicolet could correspond with isotope stages 5c and 5d. Obviously the old part of the North American record is much more obscure than the European record.

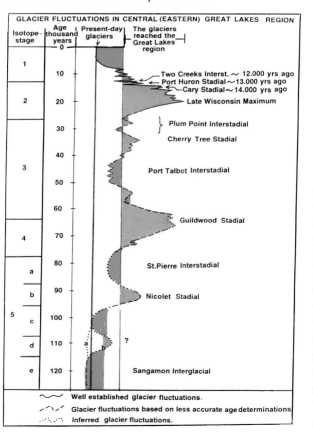

stage 3, with glaciers covering considerable parts of Fennoscandia, as evidenced by glacial deposits in northern Sweden and Finland and on the west coast of Norway. However, the climate and glaciers fluctuated, and two to five slightly warmer interstadials have been recorded in various parts of northern Europe. The Grand Pile graph indicates that there were at least five fluctuations. The interstadials were, in general, cool. During two interstadials, about 30 000 years ago and somewhere between 40 000 and 70 000 years ago, the west coast of Norway was ice free, but it is still not known how much of Scandinavia was ice free during the isotope stage 3 interstadials.

North America, 115 000–25 000 years ago (Fig. 2–18)

There are several records of Early and Middle Wisconsin climate and glacier fluctuations. However, most of them are rather incomplete, and we will focus on the Great Lakes glacial record, which currently is the best established. The record is based on radiocarbon-dated sections. However, the old radiocarbon ages tend to be considerably younger than the Uranium-series dates, which were used for the European record. Therefore, the scale in the older part has been adjusted in order to bring the start of the first cold phase back to about 115 000 years. In this way the ages of the older events are extrapolated, and the timing of the events is speculative, at best. However, the same criticisms can be raised for the timing of the old European events as well.

The correlation between the European and the American record seems striking, at least for the part younger than isotope stage 5. However, the American record for isotope stage 5 is problematic, as indicated in Fig. 2–18. The glaciers of the Glenwood Stadial, which is correlated with isotope stage 4, were almost as extensive as the isotope stage 2 glaciers. The three subsequent fluctuations (isotope stage 3?) were relatively short in distance, although the ice sheet must have been large during all of them. Unfortunately, there is no associated pollen record, such as the Grand Pile record, to document events for the Early and Middle Wisconsin times in central and eastern North America.

The large mid-latitude ice sheets during the Late Weichselian/Late Wisconsin, 25 000–10 000 years ago

General review

The graphs in Figs. 2–17, 2–18 show the general trend of Late Weichselian/Late Wisconsin climate fluctuations. The fossil flora and fauna indicate that there was a maximum cold phase from about 23 000 to about 14 000/13 000 years ago, but a considerable glacial retreat in the late part of this period suggests that some climatic warming took place at that time. Another cold phase between 11 000 and 10 000 years ago was recorded particularly in northwestern Europe. A drastic warming took place after 14 000/13 000 years ago, and the period between 13 000 and 11 000 years ago was fairly warm. A second period of drastic warming occurred about 10 000 years ago.

The ice sheets in Europe and much of North America had their maximum W/W extension about 21 000–17 000 years ago and retreated slowly, but in a fluctuating manner, between about 17 000 and 15 000 years ago. They advanced considerably during a cold peak, about 15 000 to 14 000 years ago, and in parts of North America they reached a maximum Wisconsin extension about that time (see Figs. 2–48, 2–49). The glacier retreat continued, and was relatively rapid between 13 000 and 11 000 years ago, although small glacier readvances occurred within this period also. Between 11 000 and 10 000 years ago, in Younger Dryas time, the glaciers in northwestern Europe advanced a short distance, and they generally retreated rapidly during Holocene time, after 10 000 years ago. Note that all presented ages are radiocarbon ages; see p. 184.

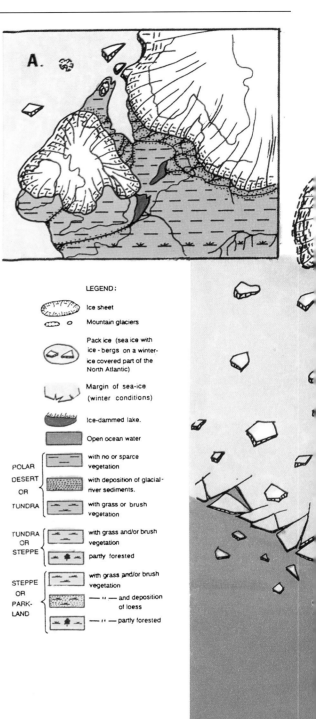

LEGEND:

Ice sheet

Mountain glaciers

Pack ice (sea ice with ice-bergs on a winter-ice covered part of the North Atlantic)

Margin of sea-ice (winter conditions)

Ice-dammed lake.

Open ocean water

POLAR DESERT OR TUNDRA — with no or sparce vegetation

with deposition of glacial-river sediments.

with grass or brush vegetation

TUNDRA OR STEPPE — with grass and/or brush vegetation

partly forested

STEPPE OR PARK-LAND — with grass and/or brush vegetation

— '' — and deposition of loess

— '' — partly forested

The maximum extent of the ice sheets, 21 000–17 000 years ago, or 15 000–14 000 years ago

Records from North America show that the ice sheets advanced in the period following 25 000 years ago, and reached the maximum limit 21 000–18 000 years ago in the eastern sections and 15 000–14 000 years ago in the western sections of the continent. The records include many radiocarbon-dated deposits related to the advances. However, the glacial advance in northern Europe is not as well recorded and dated, but the several available radiocarbon dates indicate that the maximum extent was reached sometime between 22 000 and 18 000 years ago. At the times of their maxima, the size of the North American Laurentide Ice Sheet was about 12.5 million km² and reached

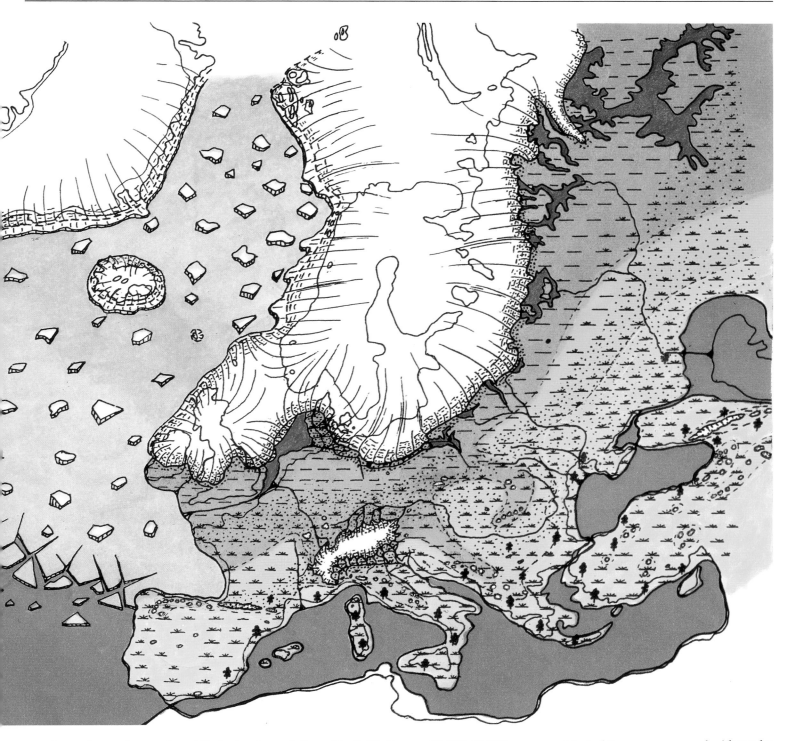

southward to about the 40° latitude, while the North European Ice Sheet was about 4.5 million km² and reached the 52° latitude (see Figs. 2–19, 2–31).

Europe

The North European Ice Sheet, while land-based in the south over Denmark, northern

Fig. 2–19. Europe, 20 000–17 000 years ago. Most of Europe was covered with tundra (orange) and open steppe or parkland (yellow). The tundra had no or very sparse vegetation in a zone adjacent to the ice sheet, and brush/grass vegetation covered much of the southern (eastern) parts of the tundra. Steppe or parkland (with patches of forest) covered much of southern Europe.

For the North Sea area the main map probably shows conditions 22 000–21 000 years ago, and the map in the small frame probably shows conditions 20 000–18 000 years ago.

The extent of the ice cover on the North Atlantic varied, and the maps show the approximate winter pack-ice conditions.

Fig. 2–20A. View of a western section of the Greenland Ice Sheet. (Photo by Höjmark Thomsen.) Note the end-moraine ridge along the ice margin and the dirt-covered marginal zone adjacent to this ridge. Sections of the margins of the North European and the North American ice sheets looked about like this. However, outwash plains with meltwater rivers were more typical for many areas beyond the margins of the ice sheets in North America and northern Europe (see Figs. 2–20B and C, and Fig. 2–21).

Fig. 2–20B. Outwash plain in front of Thorisjökull Glacier in Iceland. Outwash plains of this kind were rather common in front of the ice age ice sheets. (Photo by Inge Aarseth.)

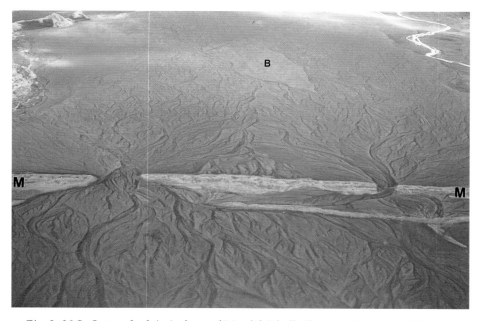

Fig. 2–20C. Outwash plain in front of Myrdalsjökull Glacier in Iceland, looking down-valley from a point above the front of the glacier. The outwash channels (now abandoned) breach a prominent young end-moraine ridge (M). Note also the small area (B), which is a "bakkeö" (hilly island) rising slightly above the outwash plain. Numerous plains of this kind were formed along the margins of the Wisconsin and Weichselian ice sheets in North America and Europe (see Fig. 2–21A). (Photo by Johannes Krüger.)

Germany, Poland, the Baltic countries, and Russia, had a slightly controversial marine-based section in the North Sea, and a marine-based section along the west and north coasts of Norway. In addition, the northernmost section merged with the Barents Sea Ice Sheet.

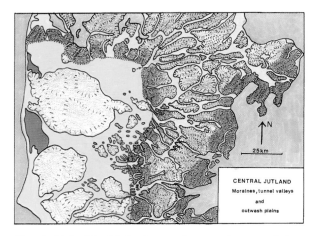

Fig. 2–21A. Glacial landscape features in central Jutland, Denmark, between the west and the east coast. (Modified from various sources, mainly from "Landskabskort over Danmark", Geografiforlaget, 5464, Brenderup.)

Dark brown: The Central Jutland and the East Jutland end-moraine zones, between 20 000 and 15 000 years old.

Lighter brown: Mainly subglacially deposited moraines and glaciofluvial sediments.

Yellow: Valleys and lowlands through which the glacial rivers drained: Tunnel valleys?

Dotted line: The maximum extent of the Late Weichselian Ice Sheet, 20 000–18 000 years ago.

Orange: Outwash plains ("hedesletter" in Danish) deposited by glacial meltwater rivers in front of the ice sheet.

Red: Low hills with old glacial/glaciofluvial etc. deposits, supposed to be of Saalian age.

Light yellow: Beach deposits.

D: The Dollerup area (see Fig. 2–22).

Note: The subglacial river drainage during the Late Weichselian maximum followed the valleys which are directed radially towards the dotted maximum line. Some of the valleys were probably eroded by the subglacial rivers, and they are true "tunnel valleys". However, many of the valleys were only partly eroded by the rivers. Valleys which are oriented approximately parallel with the dotted maximum line are "transverse valleys" (Urströmtäler) eroded by rivers which flowed along the margin of the retreating ice front.

The outwash plains were graded to a sea level, or lake level, which lay in the North Sea area far to the west of the present coast. In some areas the outwash plains were built up to a higher level than the Mid-Jutland end moraine (see Fig. 2–21B).

The southern section covered much of northern Europe, including the site of present-day Berlin, where it left a hummocky terrain with a series of end-moraine ridges, hills, and depressions which today are filled with lakes. Subglacial rivers transported large amounts of sediment, and the coarser fractions, mainly sand and gravel, were deposited in flat outwash plains in front of the ice sheet. However, before the glacial rivers reached the ice front, many of them flowed in tunnels underneath the ice sheet. They deposited sediments and formed esker ridges in some of the tunnels, but they eroded valleys (tunnel valleys) in other ice tunnels where the water flowed with high speed under high hydrostatic pressure. Several of the valleys along the east coast of Jutland and the Baltic coast of Germany were supposedly formed in this way, in part. (See Figs. 2–20 to 2–25.)

The land surface to the south of the ice margin in Germany, Poland, the Baltic countries and Russia in general slopes northwards, and therefore the rivers flowed northwards towards the ice front. In Germany most of these rivers were diverted westwards along the ice margin and eroded transverse valleys, the so-called "Urströmtäler" (Fig. 2–25). In much of Poland, the Baltic countries and Russia where the river valleys were blocked by the ice margin, many small and large ice-dammed lakes formed. The westernmost lakes in Poland drained westwards to the river Elbe, but the other lakes drained southwards in general towards the Black Sea and to the Caspian Sea, which was much larger than today (see Fig. 2–19).

In the western section along the coasts of western and northern Norway, the ice sheet flowed across the continental shelf and terminated in the ocean (Figs. 2–19, 2–30), while in the north, the North European Ice Sheet merged with a grounded ice sheet in the shallow Barents Sea area. The North Atlantic Ocean was ice covered perhaps all the way south to the coast of Portugal during the winter season. The southern part of this ice cover was pack ice, but the northern part was most likely a mixture of ice shelf and pack ice. We do not know the extent of the ice shelf. However, it certainly extended to, or slightly beyond, the outer margin of the continental shelf, as indicated in Fig. 2–19.

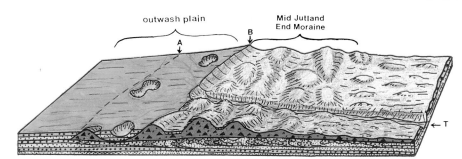

Fig. 2–21B. East-west cross section through the Mid-Jutland end-moraine zone and the corresponding outwash plain.

Orange: The outwash plain. Dark brown: The end-moraine zone. Light brown: Basal moraine and meltwater deposits. Green: Glacial deposits (till) and meltwater deposits. Yellow: Meltwater deposits. Triangles: Erratics. Large dots = gravel. Small dots = sand.

B: The maximum extent of the ice sheet when the end moraine was deposited, about 20 000–17 000 (?) years ago.

A: Suggested maximum extent of the Late Weichselian Ice Sheet about 20 000 years ago (?) T: Tunnel valley.

Note: The outwash plain was built up to a higher level than the end-moraine zone on the far side of the tunnel valley. This is true for the Dollerup area, which is shown in Fig. 2–22.

Kettle holes and till patches in the narrow zone between lines A and B indicate that the ice sheet extended slightly beyond line B before the outwash plain was formed.

The beds with meltwater sediments below the till bed are frequently folded and sheared (see Fig. 2–23).

Many long, relatively low, but fairly distinctive morainal ridges submerged on the shallow continental shelf along the coast of Norway are either grounding-line moraines or end moraines (see Fig. 2–30). The most distinctive ridges are probably true marine-deposited end moraines, and the broad, low ridges may be grounding-line moraines (see p. 126). The most distinctive ridge is the Egga Moraine along the outer margin (edge) of the continental shelf. If the reconstruction in Fig. 2–19 is correct, then this moraine could in part represent a grounding-line moraine deposited during the time of glacier maximum, when the ice shelf extended beyond the moraine. However, the Egga Moraine is most likely mainly a true end-moraine deposited during a later phase when the grounded ice sheet terminated at the moraine. Numerous seismic profiles across the continental shelf show that a thick sheet of Quaternary sediments covers the outer edge of the shelf, and the sheet wedges out towards the mainland. This sheet is stratified, and it consists of mainly glaciomarine units and till units. However, the ages of the units are not known, except that the uppermost ones are known to be of Late Weichselian age.

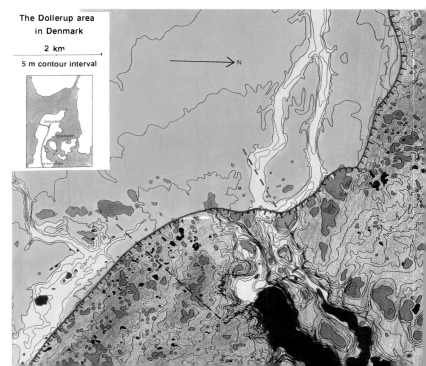

The Dollerup area
in Denmark

2 km

5 m contour interval

Legend

View shown on the panoramic picture.

The Mid-Jutland End- Moraine Zone
with ice-contact sediments.
Dark brown: hill topps.

The distal limit of the moraine

Outwash plain
(deposited by glacial rivers)

River channel on the outwash plain
formed during a late phase when the
rivers eroded rather than deposited
sediments.

Valleys and channels partly eroded by
subglacial rivers (tunnel valleys ?)

Lake

Fig. 2–22. The hummocky Mid-Jutland end moraine, about 20 000–18 000 years old, and the corresponding meltwater outwash plain at Dollerup, Denmark. The hachured line on the main picture corresponds to the hachured line on the map.

The abrupt change from the moraine with numerous hummocks and small lakes to the flat outwash plain is very striking at Dollerup. The main map and the main picture show the same area, with Lake Hald and the moraine (brown) in the foreground and the outwash plain (orange) in the background. Note the steep, hummocky ice-contact slope, about 60 m high, to the west of Lake Hald. Picture A shows a close-up of that slope with Lake Hald in the background. Picture B shows a typical glacial deposit, a till exposed on this slope. Note that the till is unsorted; the larger stones are "floating" in a finer-grained matrix. This kind of deposit is characteristic of some of the hummocky moraines. Picture C shows the flat outwash plain looking east, with the forested area of the end-moraine zone in the background. Picture D is from a gravel pit in the outwash plain. The flat-lying beds of sand and gravel were deposited by meltwater rivers, and they are typical for outwash-plain deposits.

Corresponding hummocky end moraines, deposited at the margin of the ice sheet during the Late Weichselian (20 000–18 000 years ago), can be traced continuously through Denmark, Germany, Poland, the Baltic countries, and Russia (see Fig. 2–47). Rivers from the glacier transported large amounts of sand, gravel, and stones that were deposited on the flat outwash plains which frequently border the hummocky moraines. (The main picture: photo by E.W. Olsson – Luftfoto.) (The map is modified from the topographic map of Dollerup by Geodetisk Institutt in Denmark.)

A

B

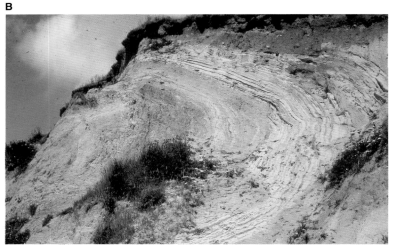

C

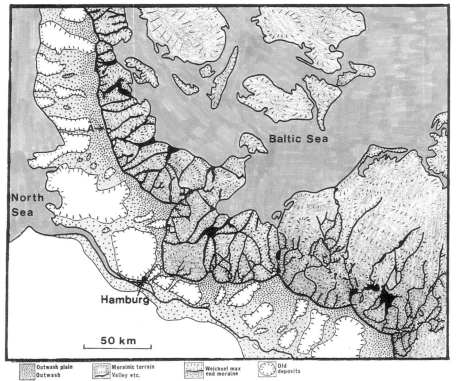

Fig. 2–23. A + B: Till beds overlying folded beds with glaciofluvial sand and gravel are common both within the Weichselian maximum end-moraine zone in Denmark and in many other areas that were covered by the Weichselian Ice Sheet. The pictures show 1–2 m thick till beds (x) deposited by glaciers which moved in the directions indicated by the arrows. They overlie stratified sediments, mainly glaciofluvial sand or gravel beds which the glaciers folded. Note that the beds in contact with the till beds are dragged in the ice-flow direction, and the fold axes are vertical to this direction. (B: Photo by Sten Sjörring.)

C: Large crystalline boulders (erratics) from the Precambrian Fennoscandian Shield are common within the till in Denmark (note that this kind of bedrock does not exist in Denmark). The farm and the fence on this picture lie in the Mid-Jutland end-moraine zone near Dollerup, and the boulders (erratics) are derived from the till.

The North Sea area has been a controversial area. Mainly on the basis of the existence of fresh morphological features on the North Sea floor, together with some radiocarbon dates, several scientists have hypothesized that the North Sea was covered with a grounded ice sheet during Late Weichselian time, and that

Fig. 2–24. Outwash plains (yellow) deposited by glacial rivers in front of the Weichselian Ice Sheet, about 20 000–17 000 years ago, and the drainage ways (black) for the corresponding subglacial rivers (modified from P. Woldstedt, 1954). Marine bays, valleys etc. are marked as drainage ways. Some of them were probably eroded by the subglacial rivers, and they are true tunnel valleys. The original outwash plains are best preserved near the Weichsel maximum end-moraine zone. The morainic terrain (brown) includes various kinds of moraines (also end moraines) and various glacial river deposits. The old deposits (white) represent, in part, glacial sediments supposed to be of Saalian age. They form low hills, and in Denmark hills of this kind, surrounded by Late Weichselian outwash, are called "bakkeöer" (hilly islands).

A: Location of the moraine and outwash plain shown in Fig. 2–26.

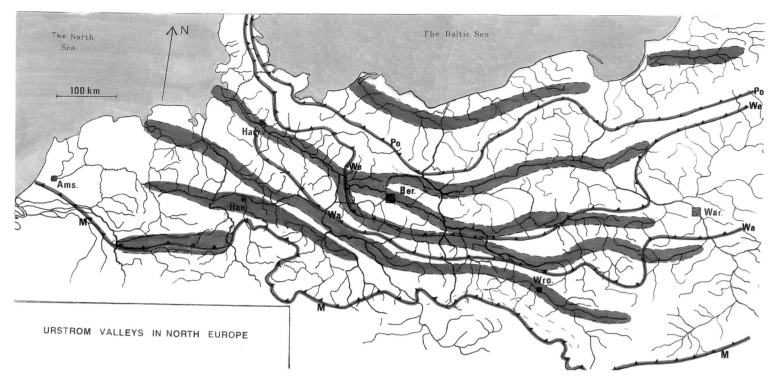

URSTROM VALLEYS IN NORTH EUROPE

here the Scandinavian Ice Sheet merged with the Devensian Ice Sheet over Britain. However, subsequent studies first led to the conclusion that there had not been contact between the two ice sheets in Late Weichselian time. Still later research suggested that the two ice sheets did merge shortly before 21 000 years ago, and separated to leave the central part of the North Sea deglaciated by about 20 000 years ago (see Fig. 2–19). This conclusion was based on detailed stratigraphic studies of radiocarbon-dated cores from the North Sea floor.

The British Isles and Ireland were covered by a small ice sheet during the Late Weichselian, presumably as indicated in Fig. 2–19. This glaciation is called Late Devensian in Britain.

Fig. 2–25. Urstrom valleys in northern Europe.
Blue belts: Main Urstrom valleys. Red lines: Main ice-margin positions during: M-M: Elster/Scale maximum; Wa: Warthe maximum; We: Weichsel maximum; Po: Pomeranean substage. Ams (Amsterdam), Ham (Hamburg), Han (Hanover), Ber (Berlin), Wro (Wroclaw), War (Warsaw).

Transverse valleys, called Urstrom valleys (Urströmtäler) in northern Europe, were formed by rivers which were diverted to flow in westerly direction along the fronts of the North European ice sheets. In addition to the valleys eroded by the transverse rivers, end moraines, accumulated at the ice fronts (on the north sides of the valleys), prevented the rivers from resuming their original northerly course when the ice sheets melted. Only the largest, best-known Urstrom valleys are marked on the map. Note how well the valleys correspond with the trend of the plotted former ice-front positions. (Modified from various sources, including E. Neef, 1970, and P. Woldstedt and K. Duphorn, 1974.)

Fig. 2–26. Distinctive Weichsel maximum end-moraine ridge (in the background) with an outwash plain in front, in Schleswig-Holstein, Germany. (Photo by Sven Christensen.)

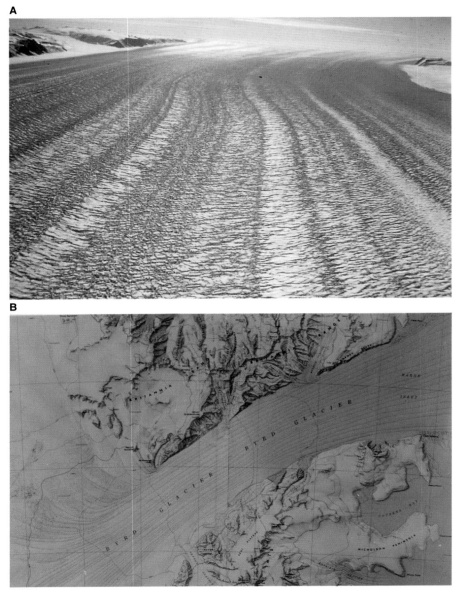

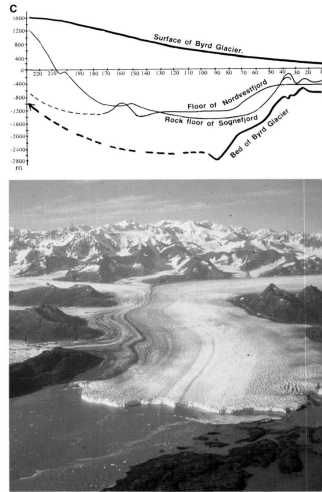

Fig. 2–28. The Columbia Glacier in southeastern Alaska. This kind of glacier occupied and took part in the erosion of the fjords mainly during periods when the climate was less severe than in Antarctica today. The Columbia Glacier has a narrow, partly floating and steep front from which small icebergs break off and float in the water in front of the glacier. Note also the greyish color of the water, which is caused by suspended clay and silt particles transported with meltwater. (Photo by Mark Meier.)

Fig. 2–27. A. Byrd Glacier, Antarctica, looking upglacier with the Polar Plateau in the background. The distance across the glacier is about 25 km. (Photo by George Denton.)

B. Byrd Glacier occupies approx. a 2800 m deep fjord/valley which crosses the Transantarctic Mountains. The glacier drains a large part of the Polar Plateau Ice Sheet and feeds into the huge Ross Ice Shelf. Glaciers of about this kind occupied and took part in the erosion of many of the deep fjords in the world, particularly in Norway, Greenland, Alaska and Chile. They existed in periods of maximum glaciation. (Published with permission from U.S. Geological Survey.)

C. Longitudinal profile of the surface and the bed of E. Byrd Glacier (thick lines). The full thick lines are modified from I.M. Whillans and others (1989). The dashed thick line: No information was obtained about this glacier bed, except that it rises to about sea level in areas beyond the left part of the illustration. Thin lines: The floors of Nordvestfjord in Greenland (the deepest partly glaciated fjord in the world), and of Sognefjord in Norway (the deepest unglaciated fjord in the world). Sognefjord has a rock floor about 1500 m deep, and Nordvestfjord is about 1459 m deep, but it has a rock floor which is most likely more than 1600 m deep.

Within areas covered by the Late Devensian Ice Sheet, the deposits generally have a "fresher" topography than areas that were not covered by this sheet. However, this difference is not always striking, and it has been problematic to determine the exact maximum limit for the Late Devensian Ice Sheet in some areas.

The Alps

The extent of the ice cover during the last glaciation, called the Würm Glaciation in the Alps, is shown in Fig. 2–7. Distinctive marginal

A

B

Fig. 2–29. A: The Fimbulisen Ice Shelf in Antarctica. This kind of ice shelf, with a high and steep ice cliff, existed along parts of the marine sectors of the large North American and North European ice sheets during the coldest periods of the late Cenozoic glaciations. Large icebergs, some with areas of several km², break off from this kind of ice shelf. (Photo by Olav Orheim.)

B: A small outlet glacier from the East-Antarctic Ice Sheet transects the Transantarctic Mountains and pushes into the Ross Sea as a narrow ice-shelf tongue. The picture was taken in January when much of the Ross Sea was open water, and only a narrow zone of sea ice existed along this part of the coast.

Parts of the marine sector of the North European and North American ice sheets were in some periods probably much like this. However, the ice shelves were obviously considerably larger and more extensive along much of the marine sectors on both continents during other periods.

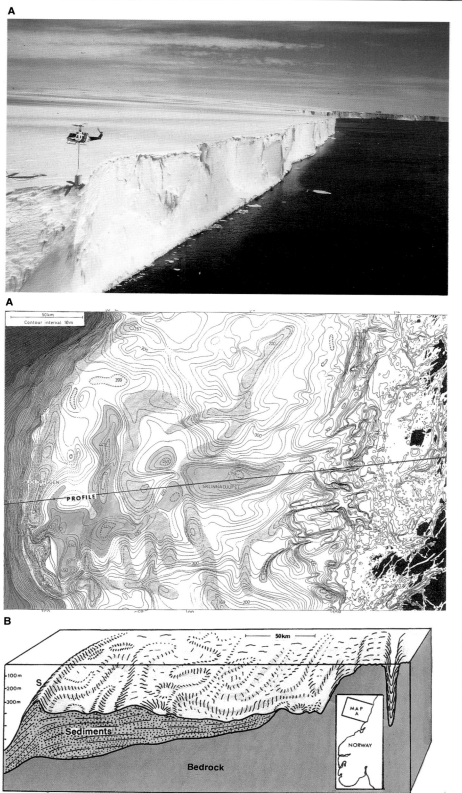

Fig. 2–30. The continental shelf off northwestern Norway is about 250 km wide and about 300 m deep at its outer (western) edge, where the large, 250 km long and maximum 150 m high Skjoldryggen end moraine (S) is located.

A: Skjoldryggen and younger end moraines or grounding-line moraines (shaded belts). 10 m contour intervals. Dark blue: 800–1000 m depth. Medium blue: 600–800 m depth. Light blue: 400–600 m depth. (Modified from O. Holtedahl, 1940.)

B: Cross section along the line x–x on map A. Yellow: Quaternary sediments, mainly glacial and glaciomarine. Black: Bedrock.

moraines were formed by the Würmian glaciers. Many of them are damming lakes, and practically all of the beautiful lakes in the Alps and in the foothills of the Alps lie in glacially sculptured basins within the limit of the Würmian glaciated areas (see Fig. 1–23). However, the lake basins must have been carved during several successive glaciations.

Numerous peaks and ridges rose above the ice surface during maximum glaciation, but the surface was high and extensive enough to support an atmospheric high-pressure area from which cold (cool) and dry katabatic winds blew downvalley into the foreland, and thus supported the creation of periglacial conditions in much of southern Europe. Together with the glaciated Carpathians and Pyrenees, the glaciated Alps formed an important barrier for the north-south migration of plants and animals.

North America

Two very extensive ice sheets covered much of North America, the Laurentide and the much smaller Cordilleran, centered on the Rocky Mountains (Fig. 2–31). Most sections of the Laurentide Ice Sheet reached their maximum extent between 22 000 and 17 000 years ago at the same time as the North European Ice Sheet. However, the Cordilleran Ice Sheet reached its maximum as late as 15 000–14 000 years ago. At about that time the Laurentide Ice Sheet experienced a readvance along parts of its southern margin, and a western lobe, the Des Moines Lobe, reached a maximum extension at that time.

At its maximum the Late Wisconsin Laurentide Ice Sheet was more than twice the size of the North European Ice Sheet. In the north it merged with, or had contact with, ice sheets over the Queen Elizabeth Islands of Canada and Greenland. In the west it was in contact with the Cordilleran Ice Sheet. Figure 2–31 presents a maximum model in which the entire eastern Canadian coast was ice covered and another rather larger ice sheet covered the Queen Elizabeth Islands. However, other scientists favor a minimum model in which the eastern Canadian coast was largely ice free and only small ice sheets covered the Canadian islands (see Fig. 2–31).

Glacial geological evidence, such as the elongation directions of drumlins and eskers (Fig. 2–70), show that at times the Laurentide Ice Sheet had two accumulation zones, or ice centers, from which the ice flow radiated: the Labrador and the Keewatin centers. The question arises as to whether the two centers existed during the ice maximum of 20 000 to 18 000 years ago, or if they developed during a later deglaciation phase. For the model reconstructions on Figures 2–31 and 2–49, two

Fig. 2–31.
North America, 20 000–18 000 years ago. Two ice sheets, a large Laurentide Ice Sheet and a smaller Cordilleran Ice Sheet, covered most of the northern part of the continent.
Forests covered much of the land area to the south of the ice sheets. Note the striking difference from conditions in Europe (Fig. 2–19). A relatively narrow tundra belt lay along the southern margin of the Laurentide Ice Sheet. The Boreal forest zone was almost absent in the east, but very broad over the southern parts of the western mountain and valley districts, where even most of present-day desert areas were forested, or partly forested. However, the western forest zone was more complex than indicated. Tundra covered parts of the highest mountains adjacent to the local glaciers, and parkland to open forest probably covered parts of some valleys.
The Laurentide Ice Sheet is presented with a Keewatin center (K) and a Labrador center (L). Scientists do not agree on the exact extent of the ice sheet over the northern Canadian islands and Baffin Bay. We have used a maximum model based mainly on information from G. Denton and T. Hughes (1981). However, many scientists favor the minimum model in the small frame (modified from A.S. Dyke and V.K. Prest, 1987). The limit of sea ice cover during winter seasons is drawn arbitrarily.

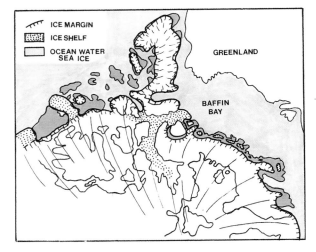

centers are used, one over Labrador and the other over Keewatin.

Rather distinctive end moraines were deposited along the southern margin of the Laurentide Ice Sheet at the maximum (about 21 000–18 000 years ago), such as the Shelbyville Moraine in the Midwest and the Ronkonkoma Moraine on Long Island, New York (Fig. 2–32A). The location of the 20 000–18 000 year old Cordilleran ice margin is not well established, but it was somewhere behind the position that the ice sheet reached at its maximum extent about 15 000 to 14 000 years ago (Fig.

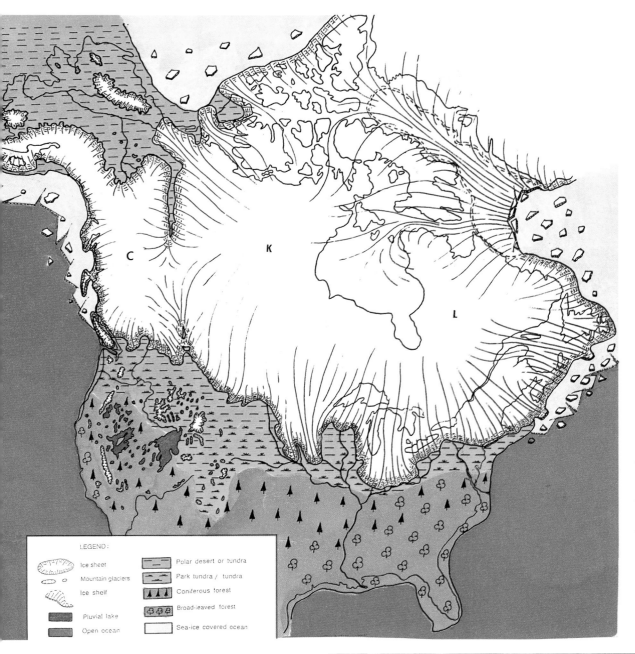

Fig. 2–32A.
The end moraines (black) between Hudson River and Cape Cod were deposited about 20 000–18 000 years ago. The dotted line represents the trend of the moraines, now below sea level. Most of the sediments on the islands and the coast are derived from the glacial deposits. (Modified from R.F. Flint, 1971.)

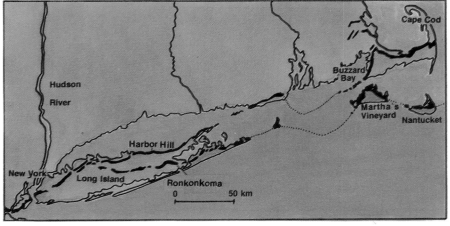

2–49). These relationships raise the question of why the Cordilleran Ice Sheet achieved a maximum extent 4000–6000 years later than most of the Laurentide Ice Sheet. The cause could be a change in the atmospheric circulation pattern which resulted in an increased precipitation in the west, about 15 000 to 14 000 years ago, and thus a late expansion of the Cordilleran Ice Sheet. This explanation seems to be supported by the contemporaneous change in vegetation, from open grassland to a more closed forest, in the western coastal districts between 15 000 and 13 000 years ago, and a shift in pluvial ac-

Fig. 2–32B. *The bouldery end-moraine zone which marks the maximum extent of the Late Wisconsin Okanogan Lobe of the Cordilleran Ice Sheet on the Waterville Plateau in Washington State. Note the hummocky morainic terrain in the area which was ice covered, and the flat outwash plains deposited by glacial rivers in front of the glacier, to the right. (Photo by John Shelton.)*

tivity from an early maximum in the eastern to a later maximum in the western Great Basin region (see p. 72). However, the end-moraine records for several local mountain glaciers both in Alaska and in the Cascade Range indicate that they had a maximum Late Wisconsin extension sometime between 22 000 and 17 000 years ago, and that the local glaciers in the Cascade Mountains experienced a considerable readvance about 15 000–14 000 years ago. This is a pattern which agrees well with the general pattern for the Laurentide Ice Sheet, but it seems less compatible with the suggested climatic change.

Another explanation for the late maximum advance of the Cordilleran Ice Sheet is focussed on topographic factors. Since most of the southern margin of the Cordilleran Ice Sheet was located on considerably higher ground and at a higher latitude than the southern margin of the mid-west section of the Laurentide Ice Sheet, the climate warming which resulted in a

retreat of the Laurentide Ice Sheet between 17 000 and 15 000 years ago had less effect on the Cordilleran Ice Sheet. Therefore the following climatic cooling, about 15 000 to 14 000 years ago, resulted in an advance of the Cordilleran Ice Sheet further than that of the eastern mid-west sector of the Laurentide Ice Sheet.

The correct explanation of the different behavior of the Cordilleran Ice Sheet will prove most likely to be a combination of climatic, glaciologic and topographic factors.

An important factor for the North American ice age climate was probably the change in the high-altitude westerly jet-stream. Modeling of the climate at 18 000–20 000 years ago indicates that this jet-stream was divided into two branches, a northerly and a southerly, and this change could have caused much of the special climate which resulted in the observed glacial and pluvial conditions.

Europe beyond the large ice sheets during the Weichselian maximum, 22 000–17 000 years ago; sea levels, climate and organic life

Low sea level

So much water had evaporated from the world's oceans and stored as ice in the large ice sheets on the continents that the world sea level dropped in the order of 100–120 m below

Fig. 2–33. *Main circulation patterns in northern Europe during the ice age maximum, and associated eolian deposits, temperature depressions and the permafrost limit.*
Several models have been presented for the ice age wind systems. The one on the map shows a very simplified model which focusses on two main elements. Deep low pressures (L) in a zone along the atmospheric Polar Front in the Atlantic generated strong westerly winds which were periodically very cold, particularly during the winter. A high pressure (H) over the ice sheet generated very strong, dry, cool to cold katabatic winds and easterly anticyclonic winds. The dominance of the two wind systems alternated.
Numbers: The Weichselian maximum depression of winter temperatures. Hachured, broken line: the southern limit of permafrost. Dark shaded: eolian cover-sand zone. Light shaded: thick loess deposits. Medium shaded: thinner and frequently more patchy loess cover.

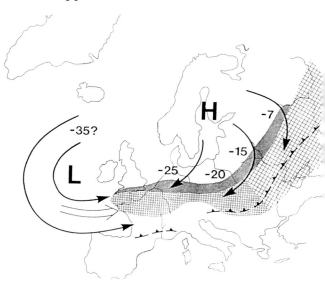

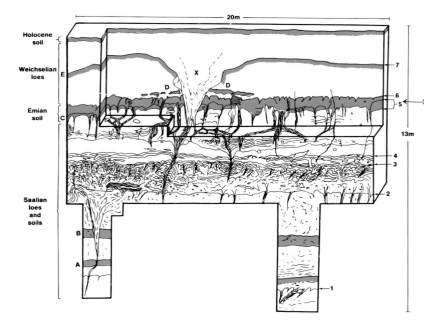

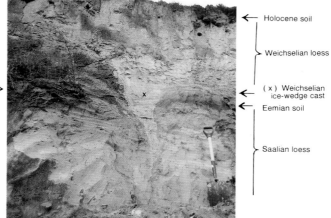

Fig. 2–34. Loess, soils and ice-wedge casts in the Gdonow section in southern Poland. The picture shows two thick loess units, a Weichselian and a Saalian, separated by a marked Eemian soil which is crossed by a large Weichselian ice-wedge cast (x). The drawing shows the entire observed Gdonow section (modified from J.M. Waga, 1987). X: The large cast shown on the picture.

A, B, C, D and E: Saalian, Eemian and Weichselian soils. The "Eemian" soil is a complex soil formed during the Eemian Interglacial, the latest part of the Saalian and the earliest part of the Weichselian. 1–7: Ice-wedge casts of Saalian and Weichselian age. Some of the weakest generations may be pseudo-casts. The Saalian beds at 3 are strongly cryoturbated.

the present. For that reason shores in southern Europe, including the Mediterranean and the Black Sea, were considerably lower than today. Large present-day submerged shelf areas were dry land, and only narrow straits at Gibraltar and Sicily separated Europe from Africa. The Mediterranean and the Black Sea were much smaller than today. However, the Caspian Sea was much larger (Fig. 2–19). There the rivers draining several glacial lakes in northern former USSR brought so much water into the lake that the lake level rose and developed an outlet to the Black Sea.

Periglacial areas

A major climatic factor responsible for the periglacial areas of Europe was undoubtedly the increased atmospheric circulation with strong, cold northeasterly (easterly) anticyclonic winds adjacent to the southeastern sector of the North European ice sheet, and strong, cold northwesterly (westerly) cyclonic winds from the predominantly ice covered North Atlantic (see Fig. 2–33). In addition the northern branch of the jet-stream over North America probably brought significant amounts of cold air into the North Atlantic area.

The cyclonic and anticyclonic circulation, in combination with the strong cold to cool, dry katabatic winds which blew from the high-

pressure areas that formed over the North European Ice Sheet and the smaller glaciated regions, in particular the Alps, created the special Arctic periglacial conditions in Europe. The katabatic winds dried up sections of the outwash plains and the soils in front of the glaciers and picked up fine-grained sediments, which were transported as dust clouds until they settled as loess over much of Europe. In fact, the extensive loess sheets in Europe (Figs. 2–34, 2–33) are the best evidence of the katabatic wind and increased circulation conditions. Many loess beds separated by soil horizons (paleosols) have been observed, with as many as seven main beds in Poland. They represent glacials or stadials, and the soil horizons represent interglacials or interstadials. Three main groups of loess beds have been recognized: (1) younger loess (from the last glaciation); (2) older loess (from the Saalian Glaciation); and (3) oldest loess (from pre-Saalian glaciations).

Katabatic winds adjacent to ice sheets are well known both from Greenland and Antarctica today. The strongest winds come in gusts, and they are usually funneled down valleys at high velocities, under the influence of gravity.

Fig. 2–35. Deposition of eolian sediments in the periglacial zones adjacent to large ice sheets. Katabatic winds in combination with strong anticyclonic (easterly) and cyclonic (westerly) winds determined most of the deposition of the eolian sediments in northern Europe.

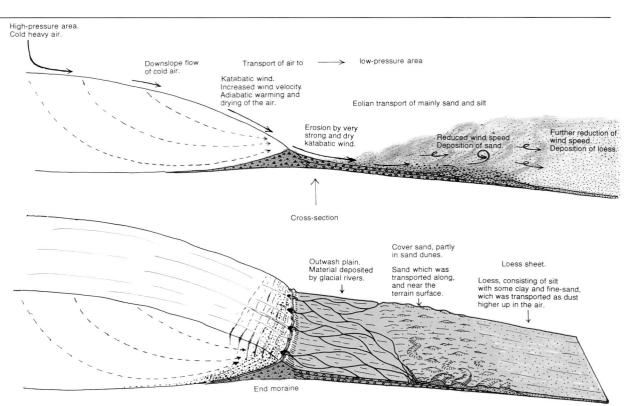

Wind velocities up to 500 km per hour have been recorded in Antarctica, and such winds can transport stones as large as cobbles. Transverse barchan-type cobble dunes have been observed in several Antarctic valleys. However, the ice age katabatic winds in Europe were apparently not so strong, but they were able, in association with the increased cyclonic and anticyclonic winds, to transport sand grains. The sand was deposited as cover sand in a wide zone closer to the ice sheet than the loess (see Figs. 2–35, 2–33). At that time the winter temperatures were approximately 20°C colder than today, and precipitation was less than half of the present in a wide zone adjacent to the ice sheet in northwestern Europe. Differences were most pronounced in the west.

Additional evidence of the severe climate over much of Europe is recorded by the presence and distribution of ice-wedge casts and other frost features which document former widespread permafrost conditions. The southern limit for these features during Late Weichselian time lies in southern France (Fig. 2–33). Today the permafrost in Eurasia is generally restricted to northern Siberia and the highest mountains of northern Scandinavia.

Still more evidence of the severe European climate is the former distribution of the vegetation zones. Tundra and open steppe/parkland-type vegetation dominated most of Europe, all the way south to the Mediterranean (Figs. 2–37, 2–19). Only small areas of Portugal and Spain, and other areas adjacent to the Mediter-

Fig. 2–36. Dust zones recorded in the ice cores from Vostok Station, Antarctica, and from Dye 3 Station, Greenland. The dust peaks correspond with the coldest glacial phases when atmospheric circulation was strongest and fierce katabatic winds were blowing.

A: The Vostok graph (modified from J. Jouzel and others, 1987). B: The Dye 3 graph (modified from C. Hammer and others, 1985). Note the distinctive isotope-stage-2 peaks and the Younger Dryas peak recorded at Dye 3.

The dust represents the finest-grained particles which were transported and spread, probably over the entire globe by jet-streams etc. The dust was generated in the way shown in Fig. 2–35, but much of it was probably generated by katabatic winds in the Tibetan Plateau region.

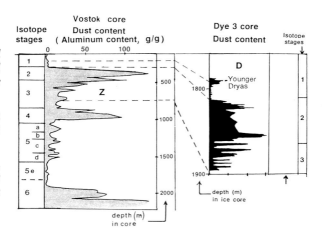

Fig. 2–37. A typical European "ice age landscape" of the kind which existed in much of Europe all the way south to the Mediterranean during the Late Weichselian maximum, and some of the Arctic animals which lived in this landscape: reindeer (A), muskox (B), mammoth (C), wooly rhinoceros (D), bison (E). Other mammals like horses and oxen were rather common also. The distribution of mammoths and wooly rhinoceroses is shown in Fig. 2–38. A similar landscape with a similar kind of fauna (except rhinoceros) existed in a narrower zone near the ice front in the eastern part of North America. In California the fauna was rather different (see Fig. 2–39).

Photos by Mittet Foto (reindeer), Reinhard Tierfoto (muskox), and paintings by Zenêk Burian and Charles R. Knight, Field Museum of Natural History, Chicago (x). The illustration of bison: from an exhibition at The American Museum of Natural History, New York.

ranean coasts, were warm enough to be partly forested.

An Arctic fauna characterized by the large animals such as reindeer, muskox, mammoth (wooly elephant), and wooly rhinoceros lived on the tundra and the steppe as far south as the Mediterranean (Fig. 2–38). Bones of these animals have been found in various caves associated with human occupations from this time, such as in the Grimaldi Caves on the Riviera coast (see Fig. 2–74 and p. 97).

The human records from many caves show that Cro-Magnon humans, with modern physical characteristics, replaced the earlier Neanderthals. The famous cave paintings from this time (Fig. 2–76), together with implements and preserved bones, show that early humans were hunters/gatherers and they hunted most of the animals mentioned above. In addition they probably hunted the cave bear, an abundant cave dweller in much of southern Europe (see Fig. 2–38).

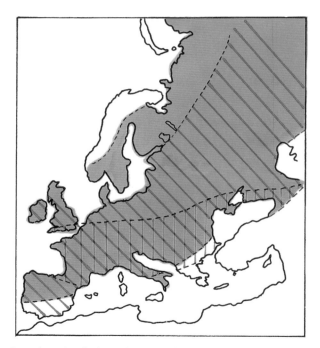

Fig. 2–38. Red: the distribution of wooly elephant (mammoth). Green shaded: the distribution of wooly rhinoceros. Red shaded: distribution of cave bear. All three animal species became extinct near the end of the Weichselian Glaciation. The distribution areas are roughly outlined, based mainly on information from B. Kurtén, 1968.
The mammoth and the rhinoceros migrated northwards in warm interglacial periods and southwards in cold glacial periods. Therefore, the northern parts of the distribution areas were inhabited during the interglacials (Eemian and Holsteinian), and the southern parts were inhabited during the cold glacials (Saalian and Weichselian). According to recent Russian observations some mammoths on a Siberian island lived as late as about 4000 years ago.

North America beyond the large ice sheets, 21 000–18 000 years ago

Meltwater rivers deposited extensive sand and gravel plains beyond the margin of the Laurentide Ice Sheet in the Great Lakes-Midwest area. They drained mainly through the Mississippi Valley to the Gulf of Mexico, and at that time the discharge of the river systems was much greater than that of today.

The climate and vegetation patterns in the periglacial areas were in some ways similar to but in several ways strikingly different from the European patterns. In general, the periglacial climate was much milder and the tundra and Subarctic vegetation zones were much narrower in North America than in Europe.

Figure 2–31 presents an approximate outline of the vegetation zones. Note that the open vegetation tundra zone was very narrow or almost non-existent in the Midwest, where an open spruce-pine forest existed close to the ice margin. However, fossil permafrost features indicate that the width of the permafrost zone could have been in the order of 200–300 km beyond the ice margin, and that most of that zone must have been forested. The open tundra/prairie belt was much wider in the east near the Atlantic coast and in the western mountain districts. An open prairie-type vegetation existed in the western coastal districts, and a rather wide conifer forest zone extended to the south of the tundra/prairie in much of the western part of the continent. The presence of a broad-leaf/hardwood forest in northern parts of the Midwest suggests that the climate here was rather warm.

Because of the wide forested areas and the narrow tundra areas, the fauna in the periglacial areas was frequently mixed. Boreal forest species were mixed with open steppe and tundra species. This is true, for example, for both mammals and beetles, which have been the most carefully studied animals and insects. The Bering Land Bridge was open during the glacials, and many of the cold-climate mammals could migrate from one continent to the other. Therefore, many of the Arctic species that lived in the periglacial areas are the same in North America and in Eurasia, such as mammoth, caribou, bison, muskox, wolves

A

B

Fig. 2–40. Loess section in Nebraska, with loess beds from two ice ages, and soils from the Sangamon Interglacial (C) and the Holocene (A). B: Wisconsin loess. D: Illinoian loess.

Fig. 2–39. Rancho-La-Brea tar pit in Los Angeles, California, contained the richest collection of fossilized Wisconsin-age mammals ever found. The animals "drowned" in the tar.

A: Painting of the ice age fauna, 20 000 years ago (section of a poster at The Natural History Museum of Los Angeles). Most of the species, such as the Imperial mammoth, the Ancient bison, the Western camel and the Harland ground-sloth, are extinct. B: Picture from present-day tar pit. Undisturbed section with fossils in the tar, exhibited in the museum.

Fig. 2–41. Loess and eolian sand in central United States. The loess sheet is thickest along the major river valleys, where most outwash from the Laurentide Ice Sheet was deposited. The loess thins gradually eastwards, reflecting the transport by predominantly westerly winds. Westwards, on the Great Plains, towards the Rocky Mountains, the eolian deposits are sandy. (Modified from C.B. Hunt, 1972.)

and lemmings, for example. Even many Boreal species, like moose, are about the same on both continents. However, the fossil species observed in California are frequently different (Fig. 2–39), and the access to this part must have been more or less closed.

Peorian loess, which is present in the Midwest and Mississippi Valley region, suggests that strong, dry katabatic winds existed. However, the distribution of the loess suggests that it was usually winds from the northwest-

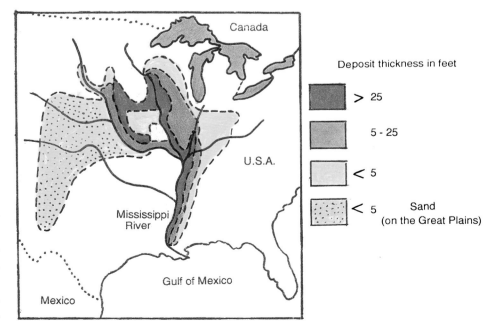

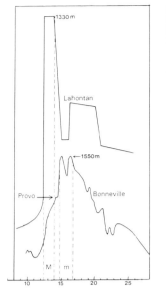

Fig. 2–42A. Fluctuations of the lake levels of pluvial lakes Lahontan and Bonneville. (Modified from L. Benson and R.S. Thompson, 1987, and D.R. Currey and C.G. Oviatt, 1985.)

M: Maximum extent of Lake Lahontan (about 14 000–12 300 years ago). m: Maximum extent of Lake Bonneville (about 16 800–14 800 years ago). Altitudes: meters above sea level.

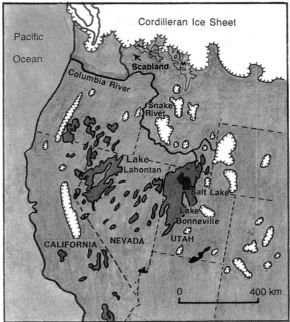

Fig. 2–42B. Principal pluvial lakes in western United States (dark blue) during the Wisconsin Glaciation. Some existing lakes are shown in black. Lake Bonneville covered an area of 50 000 km², and it was more than 330 m deep. The hachured lines define the limits of the Cordilleran Ice Sheet and Alpine glaciers. Light blue: ice-dammed lakes Columbia (C) and Missoula (M).

Fig. 2–43. Shorelines of pluvial lakes Bonneville and Provo in Provo Valley, Utah. The highest line, the Bonneville shoreline, lies about 330 m above present-day Salt Lake, and the Provo shoreline lies about 210 m above it. The lakes (see Fig. 2–42) drained across Red Rock Pass to Snake River (see Fig. 2–44). (Photo by Mary Gillam.)

Fig. 2–44. A: Red Rock Pass Spillway, looking upvalley. Lake Bonneville water spilled through this pass to Snake River. The scarp eroded by the spillwater shows particularly along the left side of the pass. The Snake River Gorge, which was partly eroded by the spillwater, lies beyond the upper right corner of the picture. (Photo by John Shelton.)

B: Snake River Gorge at Twin Falls. (Photo by John Shelton.)

ern mountain regions that caused the erosion and transport of the loess (Figs. 2–40, 2–41). The presence of forested areas close to the glacier in the Midwest indicate that the katabatic winds from the ice sheet were not extremely cold. They must have been adiabatically warmed as they flowed downslope on the ice sheet. But why the striking difference in periglacial climate and vegetation between Europe and North America? A main reason for this variation is undoutbtedly the difference in latitude of the periglacial zones on the two continents. The periglacial zone in North America lay on the same latitude as southern Italy, nearly 10° south of the European periglacial zone. Another important factor was the difference in flow of cold air from the Arctic

Ocean regions. This air could easily flow southwards across the North Atlantic and penetrate much of the European continent, as indicated in Fig. 2–33. However, in North America the Laurentide Ice Sheet was an obstacle for the southward flow of Arctic air, and only the easternmost sector of the continent was significantly influenced by this air. The rather wide open tundra/prairie zone near the east coast corresponds well with this interpretation.

Pluvial lakes (Figs. 2–42 to 2–45)

One of the most striking geographic features in the North American landscape between 18 000 and 10 000 years ago was the existence of large pluvial lakes in the present-day desert regions of the Great Valley area between the western mountain chains. The two largest lakes, Lahontan and Bonneville, covered an area larger than that of all the present Great Lakes. The eastern lakes, dominated by Lake Bonneville, attained their maximum size 18 000 to 16 000 years ago. The western lakes, including Lake Lahontan, were at their maximum between 14 000 and 12 000 years ago. A shoreline at Salt Lake City, about 1000 feet above the existing level of the modern Salt Lake, corresponds with the highest level of Lake Bonneville. Besides shorelines, extensive beds of saline lake-floor sediments were deposited within the areas covered by the extended lakes. The least saline sediments were deposited when the water level was high and the lakes were relatively well drained. Lake Bonneville had an outlet to the Snake River at Red Rock Pass. The draining of the lake took part in

eroding the spectacular Snake River Gorge (Fig. 2–44B). Some of the pluvial lakes were, in part, fed by meltwater from the glaciers, but their existence was primarily the result of reduced evaporation, combined with higher precipitation. There are many distinct shorelines at various levels below the highest ones. Studies of the shoreline sediments combined with the stratigraphy of the lake-floor sediments have revealed a history of fluctuating levels. The organic deposits within these sediments have made it possible to date the fluctuations by radiocarbon methods. The ages for Lake Bonneville and Lake Lahontan are shown in Fig. 2–42A. Note that Lake Bonneville attained a maximum size at about the same time as did the Laurentide Ice Sheet, and that Lake Lahontan attained a maximum slightly later than did the Cordilleran Ice Sheet. The close correspondence between the pluvial and glacial phases in North America is a remarkable characteristic of the ice age.

Fig. 2–45. Scabland in Washington State, USA. The spectacular canyons and channels which break the flat basaltic plain were eroded by floods due to catastrophic drainages of ice-dammed lakes about 14 000 years ago (see Fig. 2–42B). (Photo by Noel Potter.)

The melting of the large ice sheets in northern Europe and North America, 18 000–8500/8000 years ago

General remarks

The areas that were covered by the Late Weichselian/Late Wisconsin ice sheets are usually covered with deposits formed during the deglaciation period 18 000–8000 years ago. The deposits that formed during this deglaciation

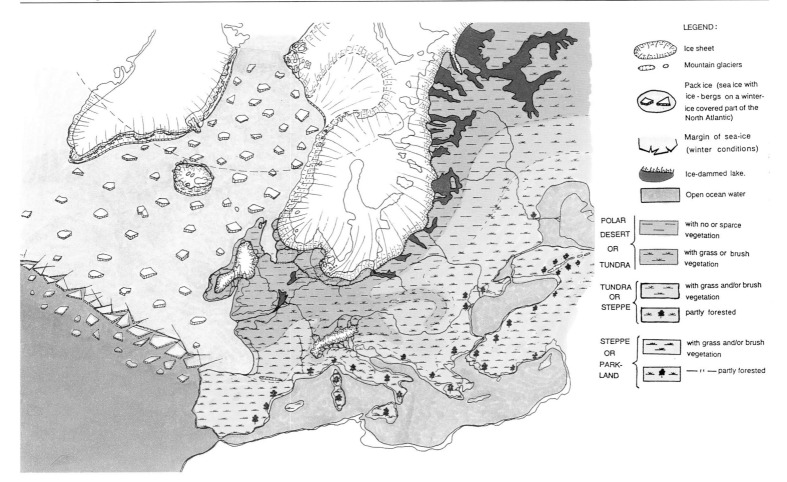

LEGEND:

Ice sheet

Mountain glaciers

Pack ice (sea ice with ice - bergs on a winter-ice covered part of the North Atlantic)

Margin of sea-ice (winter conditions)

Ice-dammed lake.

Open ocean water

POLAR DESERT OR TUNDRA — with no or sparce vegetation

— with grass or brush vegetation

TUNDRA OR STEPPE — with grass and/or brush vegetation

— partly forested

STEPPE OR PARKLAND — with grass and/or brush vegetation

— ɾɾ — partly forested

Fig. 2–46. Europe and parts of the Arctic, 15 000 years ago. The situation was much the same as 18 000–20 000 years ago, except that the ice sheets had diminished considerably. The North European Ice Sheet advanced to deposit the Main Pomeranean end moraine about this time.

are relatively very fresh and well dated. Therefore, the deglaciation history is, in general, well known. In addition, the deglaciation deposits are important to much of the present-day human activity within the regions of deglaciation.

The dominant features on both continents are as follows:

1. End-moraine complexes with numerous morainic hills and ridges, frequently containing lakes in both small and large depressions.

2. Relatively flat plains composed of basal till, sometimes with a surface morphology of streamlined hills (drumlins). Small plains of this kind lie scattered between the end-moraine complexes, while some larger plains are more extensively drumlinized.

3. Flat outwash plains which lie between and beyond the end-moraine complexes. They consist of mainly sand and gravel deposited by glacial meltwater rivers. The most extensive outwash plains commonly lie either in a zone beyond the oldest Late W/W end

moraines or in more centrally located zones related to the final deglaciation, which occurred in Holocene time after 10 000 years ago.

4. Hills (kames), ridges (eskers) and lateral ice-marginal terraces (kame terraces), composed of mainly sand and gravel deposited by glacial meltwater rivers in contact with glacier ice, are particularly common in mountainous areas that were deglaciated after 13 000 years ago, especially in areas that were deglaciated in Holocene time, when the final melting took place.

5. Glacial-lake deposits and features, including sheets of fine-grained lake-floor sediments and coarser-grained lake-shore and delta deposits are found in areas covered by the former ice-dammed lakes. Some of the largest ice-dammed lakes are shown in Figs. 2–46, 2–50, 2–53 and 2–57.

6. Glaciomarine deposits and features, such as outwash deltas and sea-floor deposits which are generally composed of clays and silts, blanket large areas in many glacio-

isostatically emerged coastal areas in both North America and northern Europe, as well as in old glaciomarine areas which are submerged today.

The deglaciation pattern

The deglaciation patterns for the North European Ice Sheet and the eastern North American Ice Sheet (the Laurentide Ice Sheet) were relatively similar, although differences did occur. The deglaciation started at about the same time on the two continents, but ended about 500 years later in North America than in northern Europe. The last remains of the ice sheet in northern Europe (in Scandinavia) disappeared about 8500 years ago, when a large ice sheet still covered much of Canada. Apparently the Laurentide Ice Sheet was so much larger than the North European that it took considerably longer time for it to melt.

Northern Europe, 18 000–13 000 years ago

The vegetation records, the fossil fauna, and the fossil permafrost features all indicate that the climate was very cold in Europe 24 000 to 13 000 years ago. The exact age of the end of this period has been questioned. In most parts of the world the coldest glacial period seems to have ended about 14 000 years ago, and a marked climate warming started shortly thereafter. But recent observations from several radiocarbon-dated stratigraphic sections in Europe indicate that the marked climate warming there started as late as about 13 000 years ago, and in some areas as late as 12 700 years ago (Fig. 2–52).

However, the main retreat of the North European Ice Sheet began about 18 000–17 000 years ago, although at slightly different times in different areas. Recent radicarbon dates from the North Sea area indicate that the retreat of the ice sheet there probably started as early as 21 000 years ago. A weak climate oscillation around 17 000 years ago, the Lascaux Interstadial, corresponds roughly with the start of the major retreat on land. The reconstruction for about 15 000 years ago in Fig. 2–46 corresponds approximately with a major glacial readvance during which the large and distinctive Main Pomeranean end moraines in Germany and Poland were deposited. The glacier overrode older glaciolacustrine sediments during this advance, but unfortunately the sediments and the readvance have not been precisely dated. The climate and vegetation to the south of the ice sheet at that time were very similar to those of about 18 000 years ago, and even the Polar Front, in the North Atlantic, was fairly close to the limit which it had 18 000 years ago. In the North Sea a considerable glacier advance seems to have occurred slightly before 15 000 years ago, and it is possible that this advance could correspond with the Pomeranean advance. There is also a possibility that a dropstone layer about 14 300 years old (a Heinrich layer: see Glossary) recorded in cores from the North Atlantic could represent a cold phase of about the same age as the Pomeranean event.

The grounding-line of the marine-based ice sheets on the shallow continental shelf of northern Europe and in the Barents Sea retreated considerably between 20 000 and 13 000 years ago. The main reason for this retreat was the combined effect of the eustatic sea-level rise (in the order of 50 m) coupled with the isostatic depression of the earth's crust on the glaciated continental shelves, caused by the load of the grounded ice sheet. This resulted in an increased water depth, calving, and increased ice retreat. However, it may seem more difficult to explain why the margin of the land-based ice sheet retreated so much during the cold period between 17 000 and 13 000 years ago. It has been suggested that the increased calving and retreat of the marine-based parts of the ice sheet resulted in a general lowering (drawdown) of the ice surface which progressively affected the land-based parts also. However, the glaciers in the Alps experienced a similar large retreat in this period, and they were not influenced by the rising ocean level. Therefore, in addition to the calving and draw-down, there must have been a climatic cause for the glacier retreat. According to the astronomic factors (the Milankovitch signals), the period after 17 000 years ago was rather favorable for ice recession due to increasing insolation. Therefore, a rise in summer temperatures was most likely a primary cause for the land-based

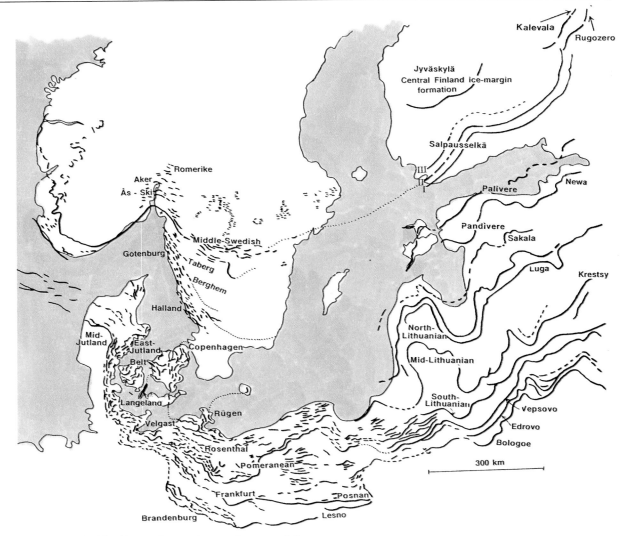

Fig. 2–47. Weichselian end moraines and some of the most used names for the moraines in northern Europe.
Heavy lines: major moraine complexes, frequently end-moraine ridges.
Dashed lines and dotted lines: connecting lines. In some cases they represent end moraines. (Modified from B. Andersen, 1981.)
Approximate radiocarbon ages in years before present: Brandenburg-Lesno: 20 000; Frankfurt-Poznan: 17 000; Pomeranean: 15 000; Mid-Lithuanean-Velgast-Copenhagen: 14 000; Luga-Rügen-Halland-Lista: 13 500; Pandivere-Gothenburg: 12 500; Berghem: 12 300; Taberg: 12 000; Ra-Middle Swedish-Salpausselkä: 11 000–10 300 (10 500); Aas-Ski: 10 500–10 200; Aker: 9800; Romerike-Jyväskylä: 9600–9500. All of the presented ages are approximate, and some of the moraines are not well dated.

glacier retreat. Apparently the glaciers were more sensitive and reacted to this climatic change more quickly than the fauna and flora.

A broad zone of the north European plain, including most of Denmark and the southwest coast of Sweden, was deglaciated during the period 18 000–13 000 years ago. Characteristic for this zone are the many end-moraine complexes composed of generally 5–20 m high end-moraine ridges, morainic hummocks, and depressions filled with lakes. Successive complexes were formed as the ice front retreated, several of them during small readvances, but the Main Pomeranean moraine complex was formed during the previously mentioned major readvance. The names of the various German complexes are shown in Fig. 2–47. Unfortunately most of them are not well dated, and the ages presented are generally based on various estimates. The prominent Main Pomeranean end-moraine complex continues eastwards into Poland, the Baltic countries and Russia, and northwards into Denmark, where it forms the hilly East-Jutland moraines. Successive transverse valleys (Urströmtäler) were also formed in northern Germany at the front of the retreating ice sheet, and ice-dammed lakes were formed in countries further to the east. Some of the lakes increased gradually in size as the ice front retreated. The size of the ice-free, dry North Sea floor also increased gradually, and the river Elbe flowed in a wide valley across this land area to a broad, shallow bay which must have been brackish and covered with sea ice much of the year. The Rhine River emptied into a lake which had its outlet through the English Channel Valley.

The size of the ice sheet over the British Isles also diminished considerably between 18 000 and 13 000 years ago, as well as the glaciers in the Alps. At the end of this period the shores in southern Britain and southern Europe were 60–80 m below the present shores. The vegetation and faunal zones were much the same as 18 000–20 000 years ago. Cro-Magnon humans who lived in southern Europe were very active hunters, and members of this group were responsible for the artistic carved bone ornaments and cave paintings which are tourist attractions today (see Fig. 2–76).

North America, 18 000–14 000 years ago

The Laurentide Ice Sheet

The period of 18 000–14 000 years ago is characterized by a general glacial retreat, but with

small fluctuations or readvances of the ice front. The most considerable readvance in the Midwest-Great Lakes region took place about 15 000–14 000 years ago in the Des Moines Lobe, which reached its Late Wisconsin maximum extent during this advance. Figure 2–48 presents the names of some of the best-known chronologically significant moraines. The southern parts of the Great Lakes basins were deglaciated around 14 000 years ago, and the first ice-dammed lakes were formed between the southern ends of the individual lake basins, such as in Lake Michigan, and the retreating ice margin (see Fig. 2–50). The vegetation pattern changed somewhat from that of 18 000 years ago, and deciduous forest trees migrated to areas close to the ice margin. However, tundra still persisted in areas northwards and eastwards along the ice margin. The fauna did not change significantly.

Fig. 2–48. Late Wisconsin-age end moraines (black) and estimated retreat positions of the ice front (hachured red lines) along the southern margin of the Laurentide Ice Sheet (modified from Dyke and Prest, 1986). All numbers on the red lines are ages in thousands of years. Large black areas and thick black lines usually represent end-moraine complexes with many hills and ridges. Most moraine segments are given local names, and only some of the most used names are listed (large numbers on the map): 1: Shelbyville (21 000–20 000); 2: Bloomington (21 000–20 000); 3: Lake Border (14 500); 4: Johnstown (21 000?); 5: Bemis (14 500–14 000); 6: Alexandria; 7: Streeter; 8: Harptree; 9: Winegar (12 000); 10: Port Huron (13 000); 11: Defiance (13 000); 12: Coba (21 000–20 000); 13: Kent (21 000–20 000); 14: Lake Escarpment (12 900); 15: Harbor Hill (20 000–19 500); 16: Ronkonkoma (21 000–20 000); 17: Nantucket (21 000–20 000); 18: Sandwich (20 000–19 500); 19: Pineo Ridge (13 500); 20: Cochrane (8200). The approximate radiocarbon ages are in parentheses.

Fig. 2–49. North America, 15 000 years ago. The Cordilleran Ice Sheet (C) was larger and the Laurentide Ice Sheet smaller than 18 000 years ago. The tundra zone had become narrower in the west and midwest, but remained extensive in the east. The maximum ice-sheet model has been used for the Canadian-islands area (see minimum model on Fig. 2–31).

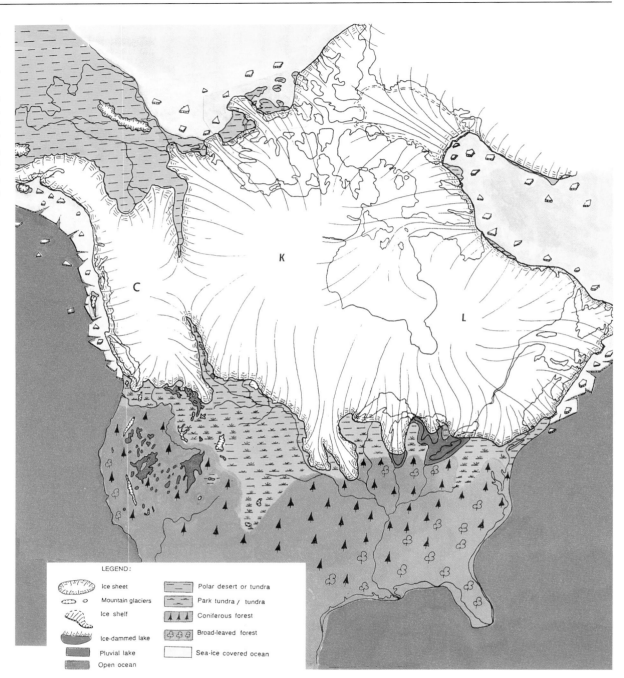

LEGEND:

Ice sheet — Polar desert or tundra

Mountain glaciers — Park tundra / tundra

Ice shelf — Coniferous forest

Ice-dammed lake — Broad-leaved forest

Pluvial lake — Sea-ice covered ocean

Open ocean

Fig. 2–50. Three selected phases during the deglaciation of the Great Lakes area. Ice-dammed lakes were formed and increased in size within the basins in front of the ice sheet as the ice front retreated. During an early phase (A) the lakes drained southwards towards the Mississippi River, while some smaller lakes drained through the Susquehanna River to the Hudson River. Later (B) the lakes Erie and Iroquois drained through an outlet at Rome, New York, to the Hudson River. During the final glacier retreat (C) the St. Lawrence Lowland was opened by deglaciation and marine water transgressed into the Lake Ontario basin. At that time the St. Lawrence area (S) was isostatically depressed more than 200 m. Since isostatic depression was greater towards the north (Fig. 1–16A), the northerly located outlets of lakes Chippewa and Stanley lay at relatively low levels, and therefore the lakes experienced a marked low-level phase. As the isostatic uplift and tilting of the land gradually progressed, the lakes were allowed to increase to the size of the present-day lakes Michigan and Superior, and the St. Lawrence Lowland emerged from the sea.

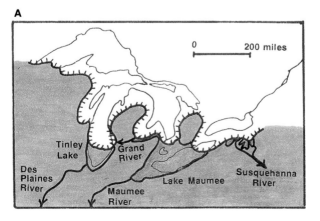

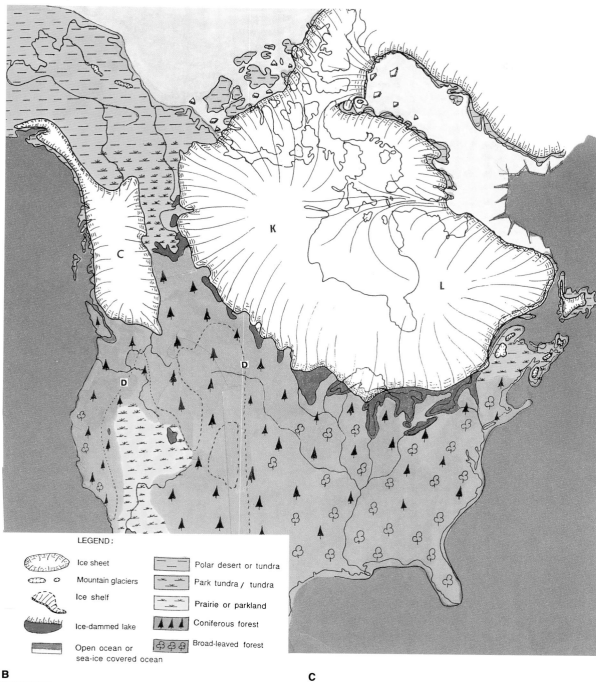

Fig. 2–51. North America, 11 500 years ago. The Laurentide and the Cordilleran ice sheets probably separated about 14 000 years ago. The mixed coniferous/broad-leaf forest zone was wide. Open woodland or parkland is indicated for many of the areas which are covered with mainly deserts today, but information about the exact conditions during the period 12 000–10 000 years ago is sparse. Present-day desert conditions were gradually established and completed about 8000 years ago, and most of the area limited by the dashed line (D) changed to prairies or deserts.

LEGEND:

- Ice sheet
- Mountain glaciers
- Ice shelf
- Ice-dammed lake
- Open ocean or sea-ice covered ocean
- Polar desert or tundra
- Park tundra / tundra
- Prairie or parkland
- Coniferous forest
- Broad-leaved forest

B

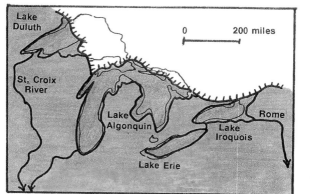

Lake Duluth
St. Croix River
Lake Algonquin
Lake Iroquois
Lake Erie
Rome
0   200 miles

C

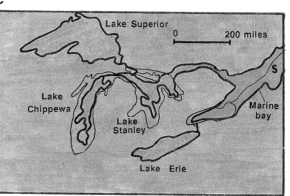

Lake Superior
Lake Chippewa
Lake Stanley
Lake Erie
Marine bay
S
0   200 miles

The Cordilleran Ice Sheet

This reached a maximum 15 000–14 000 years ago, at about the same time as the Des Moines Lobe in the Midwest (see p. 77).

Northern Europe, 13 000–11 000 years ago

The Late Weichselian warm phase; the Windermere Interstadial

The records of fossil beetles, pollen, oxygen isotopes and the marine fauna in the North Atlantic all show that a dramatic climate amelioration took place, particularly in northwestern Europe, close to 13 000 years ago. In Britain

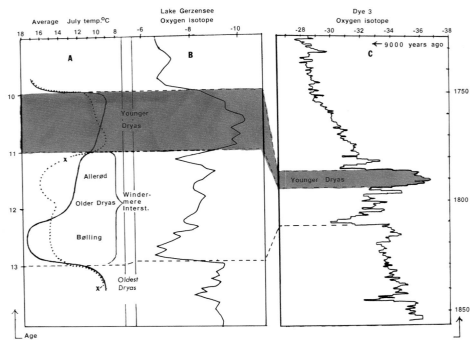

Fig. 2–52. *The rapid and drastic climate fluctuations during the late part of the Late Weichselian Glaciation are recorded with many different methods, and practically all records show the same main trends. The presented graphs represent some of them.*

A: *The heavy line, which shows the average July temperature fluctuations in Great Britain, is based on observed fossil beetle faunas (modified from G.R. Coope, 1977).*
The dotted line (x) shows the approximate trend of climate fluctuations in the area including the Netherlands, northern Germany and southern Scandinavia, based on vegetation changes. B: The oxygen-isotope, δ¹⁸O (‰), fluctuations recorded in cores from Lake Gerzensee in Switzerland. (modified from Siegenthaler and others, 1984). C: The oxygen-isotope, δ¹⁸O (‰), fluctuations recorded in the Dye 3 ice core, southwestern Greenland (modified from S. Paterson and C. Hammer, 1987). Graphs almost identical with the Dye 3 graph have now been constructed on the basis of observations in cores from central Greenland (Summit), northwest Greenland (Camp Century), and eastern Greenland (Renland).

the summer temperature increased in the order of 8–10°C, and as a consequence the glaciers retreated rapidly. In Britain the Devensian Ice Sheet probably disappeared completely by about 13 000 years ago, and a limited amount of "warm" Gulf Stream water had begun to enter the North Atlantic and the North Sea area. In southern Europe the marked warming started earlier than 13 000 years ago.

The warming was the start of the recently defined Windermere Interstadial, which lasted from about 13 000 years ago to about 11 000 years ago, and it includes the traditional Bölling, Older Dryas and Alleröd. The beetle-fauna record from Britain indicates that the early part of the period was the warmest, and that the climate "gradually" cooled during the later part. This correlates very well with the available oxygen-isotope records. However, in much of Europe to the east of Britain, the later part, the Alleröd, seems to have been as warm as or possibly warmer than the early part.

The North European Ice Sheet retreated considerably in this period, but it also fluctuated, and several smaller, recessional end-moraine complexes were deposited along the north coast of Germany, Poland and the Baltic countries, as well as in southern Sweden and southern Norway (deposited between 13 500 and 11 000 years ago; see Fig. 2–47).

An ice-dammed lake, the Baltic Ice Lake, began as a small lake in the southern part of the Baltic Sea basin about 13 000 (13 500) years ago, and it increased in size as the ice front retreated to reach a maximum nearly 11 000 years ago (see Fig. 2–53). Baltic Ice Lake sediments, frequently varved clays, cover considerable areas in coastal districts where the ice lake was more extensive than is the present Baltic Sea.

Figure 2–53 presents a general picture of the conditions 11 500 years ago, when the Baltic Ice Lake was near its maximum. At this time, the English Channel and much of the North Sea were still dry land, and a marine bay occupied Land Skagerrak the Norwegian Channel. Much of the North Atlantic was open water, at least during the summer. The forest zones, which had started a rapid migration northwards about 13 000 years ago, reached a position where much of Europe was forested by about 11 500 years ago. However, there was still a broad zone of open steppe/parkland at

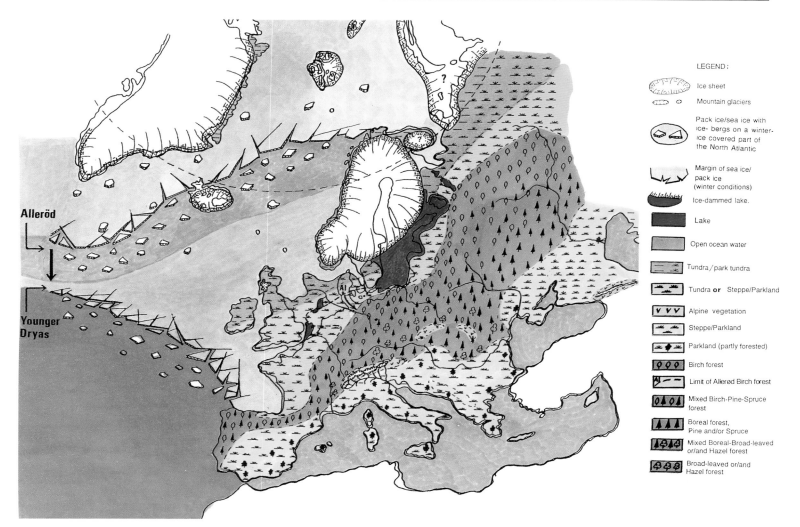

Alleröd

Younger Dryas

that time, with a tundra zone still closer to the receding ice margin. The Arctic fauna, including reindeer, muskox and mammoth, was forced to migrate northwards to live in this zone, and was replaced to the south by forest-dwelling animals. Early humans, who seem to have favored reindeer hunting, also migrated northwards with the animals (see p. 96).

Fig. 2–53. Europe and parts of the Arctic, 11 000–10 000 years ago during the Younger Dryas period. A considerable lowering of the temperature in early Younger Dryas time resulted in an expansion of the tundra, a small expansion of the ice sheet, and a considerable expansion of the pack ice on the Atlantic Ocean. The map shows the approximate limits of the winter ice cover during both the Alleröd (11 800–11 000 years ago) and the Younger Dryas (11 000–10 000) periods. The Fennoscandian Ice Sheet was only slightly smaller, and the northern European forest-zone boundaries lay slightly further to the north during the Alleröd than during the Younger Dryas period. The marked dashed line represents the Alleröd forest limit in northwestern Europe. The Baltic Ice Lake was ice covered during much of the year. Information about the glacial conditions in northernmost Russia is conflicting.

North America, 14 000–11 000 years ago (Fig. 2–51)

A significant climatic amelioration and an associated retreat of the ice front took place in all sectors of the North American ice sheets after 14 000 years ago. The coast of Maine was deglaciated, and stratified marine end moraines and ice-front deltas, together with a series of extensive eskers, were deposited. The Cor-

dilleran Ice Sheet retreated considerably, and a wide corridor opened between the Cordilleran and the Laurentide ice sheets. Most of the Great Lakes region was gradually deglaciated, and a series of ice-dammed lakes were formed in the lake basins in front of the retreating glacier (see Fig. 2–50). These lakes drained first southwards across the water divides to the Mississippi River and the Gulf of Mexico, later to the Hudson River, and finally to the St. Lawrence River into the North Atlantic.

Many end moraines were deposited during small readvances of the ice front during the overall retreat (see Fig. 2–48).

The vegetation pattern changed, but not drastically. Deciduous forest trees spread to larger areas adjacent to the ice sheet, but open

tundra/steppe persisted in some of the deglaciated eastern coastal regions. The katabatic winds diminished, and almost no loess was deposited in the Midwest after about 14 000 years ago.

With the opening of the corridor between the Laurentide and the Cordilleran ice sheets, a route became available for the first early humans to migrate into most of the New World. These people probably hunted and followed the herds of big game animals like caribou, bison, and mammoth through the so-called ice-free corridor. The oldest unquestionable evidence of early humans in the New World is from 14 000–12 000 years ago; numerous dated sites with human remains are from later times.

Fig. 2–54. A Younger Dryas, Salpausselkä II, end moraine at Vääksy in Finland, looking west (photo by Ari Lyytikäinen). The Salpausselkä II ridge separates Lake Päijänne (right) from Lake Vesijärvi (left). The ridge was deposited at an ice front which lay in the Baltic Ice Lake, and it consists of predominantly stratified meltwater river sand and gravel. Numerous gravel pits (x) lie within the Salpausselkä I, II and III ridges of Younger Dryas age. They represent a most important gravel source in Finland.
Yonger Dryas ice-front deposits, including end moraines and ice-front deltas, are very dominant in many parts of Fennoscandia.

Fig. 2–55. An end-moraine ridge within the Kalevala end-moraine zone in Russia, near Finland. The ice lay on the left side of the ridge, and meltwater rivers deposited a large outwash sand and gravel plain on the right side. The ridge is about 4–8 m high, but Kalevala ridges up to 40–60 m high have been observed in other parts. The Kalevala moraine zone has been traced fairly continuously for more than 400 km from the Finnish border. It corresponds to the Salpausselkä II and/or III moraines in Finland (see Fig. 2–47).

The Younger Dryas chronozone, 11 000–10 000 years ago (Fig. 2–53)

A drastic cold phase in north-western Europe

Fossils and the oxygen-isotope records show that the warming trend reversed and the temperatures dropped considerably in much of Europe about 11 000 years ago. The temperature's lowering was most drastic in northwestern Europe, where the drop in summer temperature was in the order of 8–10°C, according to some calculations. As a consequence glaciers in western Scandinavia advanced, and

a new ice sheet, the Loch Lomond, formed over the Highlands of Scotland. At the same time the Oceanic Polar Front in the North Atlantic pushed southwards to reach a position fairly close to the one achieved 18 000 years ago, leaving most of the North Atlantic pack ice covered during the winter months. This period is today described as the Younger Dryas cold period, since fossil Dryas plants are commonly found in terrestrial Younger Dryas sediments (see Fig. 2–68).

A Younger Dryas cold phase and an associated glacier readvance were recorded in the Alps and in Siberia too, and recent research indicates that glacier advances of about this age occurred in several other parts of the world.

The Fennoscandian end moraines (Figs. 2–54 to 2–56)

The glaciers advanced as much as 30–40 km in parts of western Norway during Younger Dryas time, but the advance was considerably smaller in eastern Fennoscandia. The fossil records indicate that the earliest part, 11 000–10 500 years ago, was the coldest, and the glacier advance culminated rather early in this period in much of Fennoscandia.

Figure 2–53 illustrates some key physical conditions in Europe and the North Atlantic during the Younger Dryas. Very prominent end-moraine ridges were formed in Europe along much of the Younger Dryas ice front. An almost continuous belt of ridges has been traced through the fjord districts of northern, western, and southern Norway (the Ra moraines), and through central Sweden (the central Swedish moraines), southern Finland (the Salpausselkä moraines) and the Karelian part of former USSR (the Kalevala morainal belt). The Loch Lomond Ice Sheet in Scotland advanced to deposit the Loch Lomond end moraines, and glaciers in the Alps – and probably in Iceland and in Greenland too – advanced to deposit marked end moraines. The North Sea land area was still extensive, and the Baltic Ice Lake was at its maximum and final stage. The northward migration of the forest belts in Europe was halted.

Forest limits were pushed somewhat southwards, and there was an increase in the area of the tundra belt, which then covered consid-

A

B

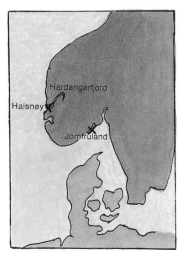

Fig. 2–56. Younger Dryas end moraines at Jomfruland Island (A) on the southeast coast of Norway and on Halsnöy Island (B) at the mouth of Hardangerfjord on the west coast. The crest of the moraine ridge rises above sea level and forms Jomfruland Island. The farmland on Halsnöy Island lies on the moraine ridge and the corresponding glaciomarine deposits. Observe that bare bedrock is exposed on the islands and the mainland on the proximal side, behind the ridges. (Photos by Fjellanger-Wideröe A/S.)

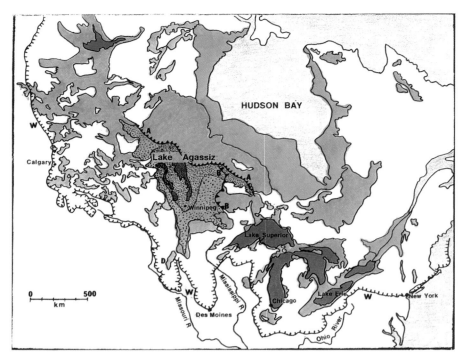

Fig. 2–57. Former ice-marginal lakes (blue) dammed by the Laurentide Ice Sheet during the deglaciation of North America, and areas submerged below ocean water (green) during the deglaciation, 13 000–8000 years ago. Submerged areas on most of the east coast of USA and Canada are not shown.

Dark blue: present-day lakes. Dotted blue: Lake Agassiz about 10 400 years ago (densely dotted) and about 9900 years ago.

A-A: the ice margin 10 400 years ago. B-B: the ice margin 9900 years ago. Hachured line (W-W): Wisconsin maximum ice margin.

D: Glacial Lake Dakota (see Fig. 2–59). Emergent glaciolacustrine/glaciomarine sediments cover most of the light blue and green areas. (Modified from J. Teller, 1957.)

Younger Dryas end moraines have been found in Greenland. Therefore, it seems rather evident that this cold phase must have influenced at least the adjacent coast of North America, and indeed it has. Recent studies have shown that a Younger Dryas cooling and glacier advance occurred in Newfoundland and Nova Scotia, while the pollen records both in Nova Scotia and in the western part of the Midwest region of the United States show evidence of a Younger Dryas cooling.

Younger Dryas in other parts of the world

Records from most of the other parts of the world are generally not so detailed that a Younger Dryas cold phase can be clearly distinguished, although in some areas scientists claim to have recorded both Younger Dryas end moraines and vegetation changes. For instance, end moraines in New Zealand are of Younger Dryas age.

What caused the Younger Dryas cold climate?

The Younger Dryas cold climate, reaching full glacial conditions once again, lasted for a maximum of 1000 years. Therefore, it was a short climatic fluctuation. But what was its cause? How could the climate in northwestern Europe change so rapidly from nearly full interglacial conditions to full glacial conditions about 11 000 years ago? This happened during a period when the orbital signals strongly favored warm conditions. Several theories have been proposed, such as:

1. A collapse of the Barents Sea Ice Sheet resulted in considerable transport of icebergs to the North Atlantic and consequently triggered a cooling and freezing of the surface waters.

2. The drainage of cold meltwater from the Great Lakes, largely through the marine St. Lawrence Lowland to the North Atlantic, occurred at about this time, and it resulted in a diversion of the important Gulf Stream and a cover of cold "fresh" water over much of the North Atlantic, a cover which froze and influenced the oceanic and atmospheric circulation.

erable parts of northernmost Europe. Coastal areas that had been deglaciated before 11 000 years ago were still isostatically depressed. Therefore, the shore level was higher than at present in much of Fennoscandia and in Scotland (see Fig. 1–15). However, the global sea level was in the order of 50 m lower than today, and therefore the shores in southern Europe, including the Mediterranean and the Black Sea, were actually about 50 m lower than today.

Younger Dryas in North America

There is no very clear evidence of a Younger Dryas cold phase and glacier advance in much of North America. However, the evidence of a Younger Dryas cold phase is quite marked in the ice-core records from Greenland and the Canadian islands (Fig. 2–52), and supposed

3. Changes in insolation caused by volcanic ash in the atmosphere (ash from volcanic eruptions mainly located in Iceland).
4. Changes in the pattern of the atmospheric jet-streams.
5. Changes in solar radiation.
6. Strong winds caused by increased high-pressure conditions over Greenland and the Arctic Ocean blew cold surface water and pack ice southwards to cover much of the North Atlantic, and thus changed the climate.
7. Changes in the interaction between ocean currents and atmospheric circulation.
8. Fluctuation in the atmospheric carbon-dioxide content, causing atmospheric cooling.
9. A surge of the Canadian Ice Sheet.

Other theories have also been suggested, but objections have been presented against all of them, and at present it seems difficult to put forward one that is preferred. At one stage many scientists favored proposition 2, but today some of the other alternatives seem to be as much in focus, such as propositions 7 to 9.

The rapid and drastic climatic change which occurred about 11 000 years ago was followed by a rapid change back to warm interglacial conditions about 10 000 years ago, and recent observations in ice cores from Greenland also show that several drastic climatic changes occurred very quickly. This shows that the stability of the world climate is vulnerable, and it raises the question: Can this kind of quick change happen again in the near future?

Lake Agassiz and the opening of the St. Lawrence meltwater drainage-way (Figs. 2–57 to 2–59)

Glacial Lake Agassiz started forming as an ice-dammed lake in the Winnipeg, Manitoba, area around 12 500 years ago, and it reached a maximum about the time of the Younger Dryas, when it was allowed to drain eastwards to the Great Lakes as the glacier margin retreated northwards. At its maximum the lake was larger than the combined present area of the Great Lakes. At about this time a drastic change in the drainage of the ice-dammed Great Lakes occurred. The ice front had retreated so much

Fig. 2–58. The abandoned early spillway of glacial Lake Agassiz on the South Dakota-Minnesota boundary, now partially occupied by Big Stone Lake, looking southeastward, downstream. (Photo by John Shelton.)

Fig. 2–59. The emerged flat floor of glacial Lake Dakota near Aberdeen in South Dakota (see Fig. 2–57) is covered with fine-grained glaciolacustrine sediments. Much of the good farmland in northern North America lies on this kind of sediment. (Photo by John Shelton.)

that a passage eastwards through the St. Lawrence Lowland was opened, and the drainage of all lakes shifted from the Mississippi Valley to this new drainage direction (see Fig. 2–50). It has been suggested that this change, in combination with the drainage of Lake Agassiz, led to a near-catastrophic outflow of cold fresh water to the North Atlantic, which could have

Fig. 2–60. The Preboreal (about 9500 years old) Odda end moraine at the head of Har-dangerfjord in western Norway (see Fig. 2–56). The town of Odda lies on the moraine and corresponding marine deposits. Sörfjord, a branch of Hardangerfjord, lies in the background, Lake Sandvinvann in the foreground, and the small Folgefonnen Glacier, a plateau glacier, in the left background. The moraine ridge Trollgaren, shown on Fig. 1–7, probably corresponds in age with the Odda moraine, and similar moraine ridges of this age lie on the plateau on the right side of Sörfjord. (Photo by Fjellanger-Wideröe A/S.)

initiated the Younger Dryas cold phase. For a short period, about 9900 years ago, the east-ward drainage of Lake Agassiz was again blocked by an expansion of the ice sheet (see Fig. 2–57).

The warming and final deglaciation, 10 000–8000 years ago

Early Holocene time

Beginning about 10 000 (10 400) years ago the climate in North America and in Europe warmed significantly. This marked the start of the Holocene Epoch, and resulted in a rapid retreat of the glaciers and an expansion of the forested areas. However, the active gla-ciers still fluctuated and deposited marginal moraines. The youngest marginal moraines were deposited about 9200 years ago in Scandinavia and about 8200 years ago in North America (the Cochrane Moraine). The last

remains of the ice sheet disappeared about 8500 and 8000 years ago, respectively, in Scandinavia and Canada.

Europe after 10 000 years ago, in Holocene time

Preboreal (10 000–9000 years ago) and early Boreal time

The North European Ice Sheet retreated rapid-ly after 10 000 years ago, and by 9500 years ago the situation was approximately as shown in Fig. 2–66. At about this time the glacier ad-vanced (or the retreat halted), and prominent end moraines or ice-contact outwash terraces were deposited along much of the ice margin. Even some older and younger, Preboreal-age marginal moraines and outwash terraces were deposited in the fjords and valleys on the coast of Norway (Fig. 2–60). Outwash terraces and a series of small end moraines, frequently called "annual" moraines, were deposited in parts of Sweden, Finland and Norway. However, the most dominant features from this period, and from the final melting phase of the North European Ice Sheet, 9200–8500 years ago, are the large esker systems, particularly in Sweden and Finland (Fig. 2–61). In addition various deglaciation-associated features, including kame terraces, hummocky dead-ice accumula-tions (Fig. 2–62), and channels eroded by gla-cial rivers, are common in the central Scandi-

Fig. 2–61. Esker ridges deposited by meltwater rivers during the melting phase of the Fennoscandian Ice Sheet are particularly dominant in Sweden and Finland. They are generally winding ridges on the deepest part of the valley floors. Most of them were deposited in subglacial tunnels. The eskers consist mainly of stratified sand and gravel, and are important sources for building material.

A: The map shows some of the main esker ridges. (Modi-fied from "Map of Quaternary Deposits of Finland, 1984" by the Geological Survey of Finland, and from E. Granlund and G. Lundqvist, 1949.)

B: The Rörström Esker in central Sweden. (Photo by Rolf Åke Larsson.)

C: The Punkaharju Esker in central Finland. Note also the numerous lakes in the background. (Photo by SK Foto.)

D: Close-up of the Punkaharju Esker. (Photo by Markku Varjo.)

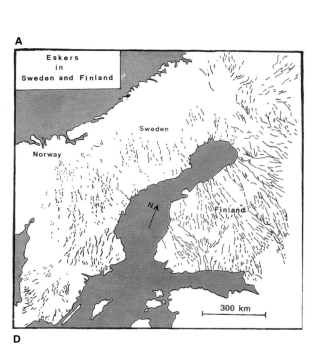

A

B

C

D

Fig. 2–62. A: Kame terraces and dead-ice topography on the south side of Döraalen Valley, Rondane Mountains, central Norway. Most of the sediments were transported by glacial meltwater rivers from Rondvass Valley (R), and deposited partly on the surface of the glacier in Döraalen Valley (resulting in dead-ice topography) and partly along the sides of the glacier (resulting in kame terraces).

The lowest-lying terraces (white path), close to river level, are postglacial river terraces. Kame terraces and areas with dead-ice topography are rather common in the central part of Scandinavia and parts of North America, which were deglaciated in warm Holocene periods when the glaciers were climatically dead.

B: *The kame terraces in Dörålen Valley, looking upvalley. The picture was taken from a kame terrace where the cobbly-bouldery outwash gravel is exposed.*

C: *Close-up of the dead-ice topography with kettle holes from central part of Fig. 2–62A. Glaciofluvial cobble-boulder gravel dominates the sediments, but patches of ablation till are common also.*

B   **C**

Fig. 2–63. A: A glacial lake in Iceland. (Photo by Eivind Heder.) B: Ice-dammed mountain lake. During the final melting phase of the North European Ice Sheet, between 9000 and 8500 years ago, a great number of small lakes were dammed by the ice in the mountain valleys of Fennoscandia (see Fig. 2–67). The lakes were formed when the ice in the upvalley parts, near the water divides, melted, and thicker ice still remained in the downvalley parts, i.e. the ice divide was located downvalley, generally to the east of the water divide. Several of the lakes drained suddenly, and the deep canyons were eroded by the floods (see Fig. 2–65). Green: lake sediments.

navian mountain districts, and lakes of various sizes were dammed by the waning ice in several high-mountain valleys (Figs. 2–63, 2–67). The upstream parts of these valleys were deglaciated before the downstream parts,

where the ice was thicker. In general the ice divide was located to the east or south of the water divides over which the ice-dammed lakes drained. However, being dammed by ice, the lakes were unstable, and they frequently drained through or below the ice in catastrophic fashion. The enormous water volumes that were discharged rapidly in this way eroded several deep canyons. The Jutulhogget Canyon in central Norway is a good example (see Figs. 2–64, 2–65). In Sweden, the "Central Jämtland Ice Lake" drained catastrophically about 8800 years ago. Evidence of this drainage was found in the form of a thick sediment bed within the Swedish varved sediments, and De Geer used this thick marker bed as a zero-year in his varve chronology (see p. 150).

Most of the North Atlantic rapidly became ice free after 10 000 years ago, and a Boreal marine fauna migrated into the North Sea and along the coasts of western and northern Norway. The sea transgressed southwards across the dry part of the present North Sea floor, but the southern part of the present-day North Sea floor still remained dry land at this time, and the marine connection through the English

Fig. 2–64. Jutulhogget ("the giant's cut") in central Norway was eroded during a catastrophic drainage of ice-dammed Lake Glomsjö about 8800 years ago. Today Jutulhogget is a dry canyon which crosses a mountain ridge between Glomdal Valley, in the foreground, and Tylldal Valley, in the background.
According to a legend the giant in Glomdal made the cut in an attempt to divert the river from Glomdal to Tylldal, in order to drown the Tylldal giant. However, he was killed before he finished the cut. Much of the rock material which was quarried from Jutulhogget was deposited in a broad ridge (R) in Tylldal Valley. Stones as large as small houses lie on the surface of this ridge. (Photo by Fjellanger-Wideröe A/S.)

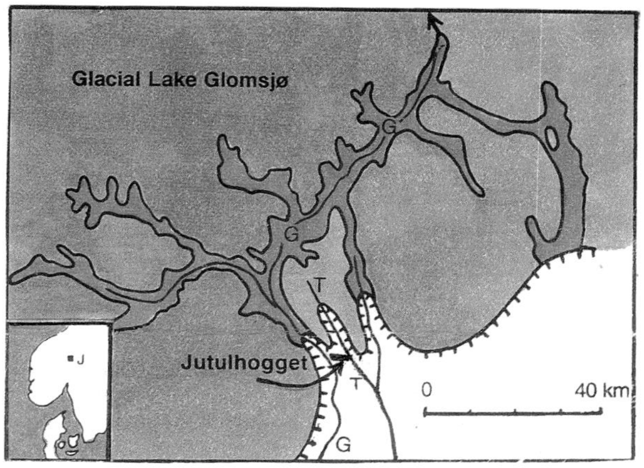
Glacial Lake Glomsjø
Jutulhogget
0
40 km

Channel was not established until about 8000–8500 years ago.

The deglaciated Fennoscandian coastal districts were in general still so isostatically depressed that the shorelines which were formed in this time were subsequently uplifted to altitudes well above the present shores (see Fig. 1–15).

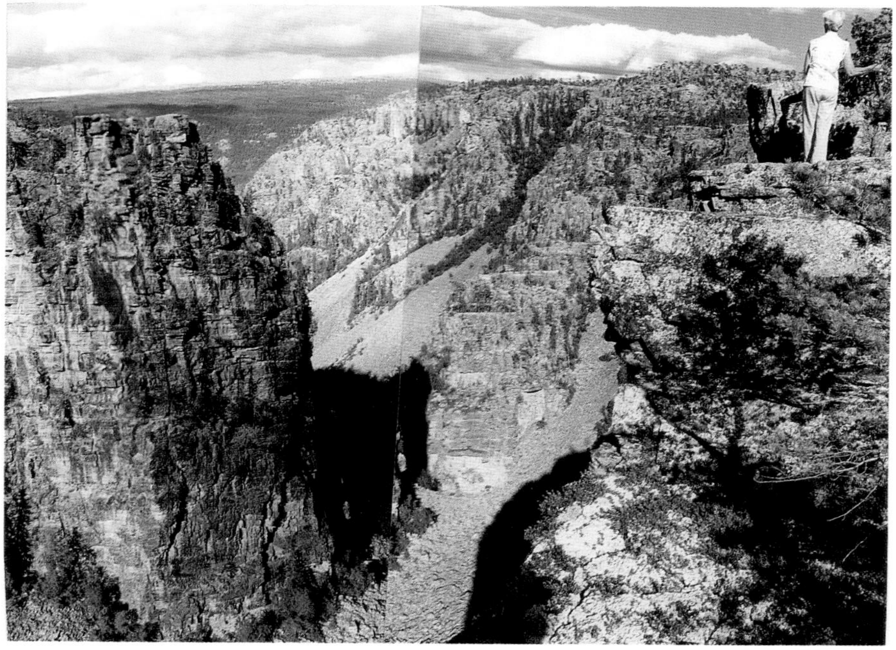

Fig. 2–65. Jutulhogget.

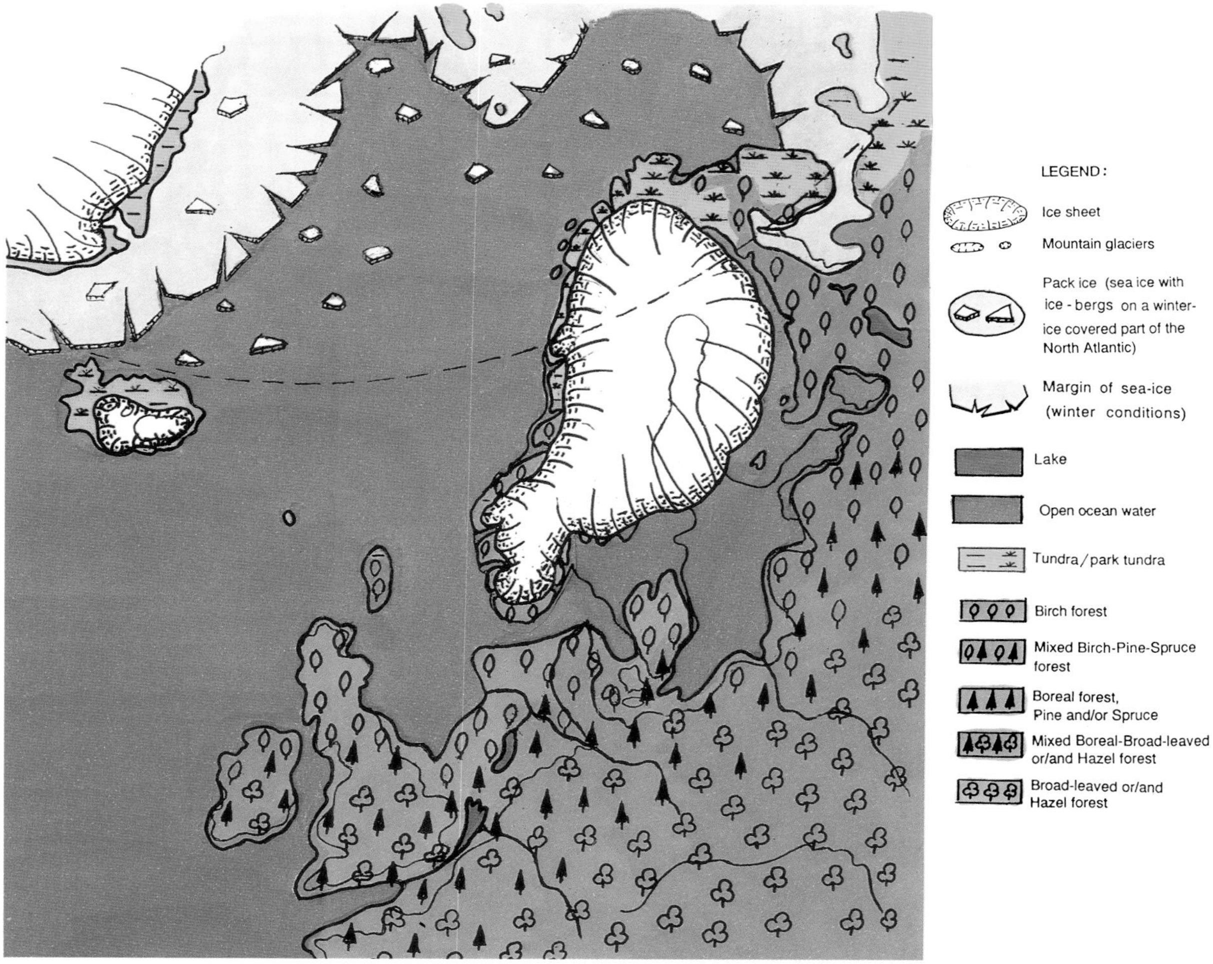
LEGEND:
Ice sheet
Mountain glaciers
Pack ice (sea ice with ice - bergs on a winter-ice covered part of the North Atlantic)
Margin of sea-ice (winter conditions)
Lake
Open ocean water
Tundra / park tundra
Birch forest
Mixed Birch-Pine-Spruce forest
Boreal forest, Pine and/or Spruce
Mixed Boreal-Broad-leaved or/and Hazel forest
Broad-leaved or/and Hazel forest

Fig. 2–66. Northwestern Europe, 9500 years ago. The Baltic Sea experienced a marine (salt-water) phase, "The Yoldia Sea". The youngest well-defined marginal moraines and ice-contact deltas were deposited in Fennoscandia about this time. Most of Europe was forested. Information about conditions in northernmost Russia is conflicting.

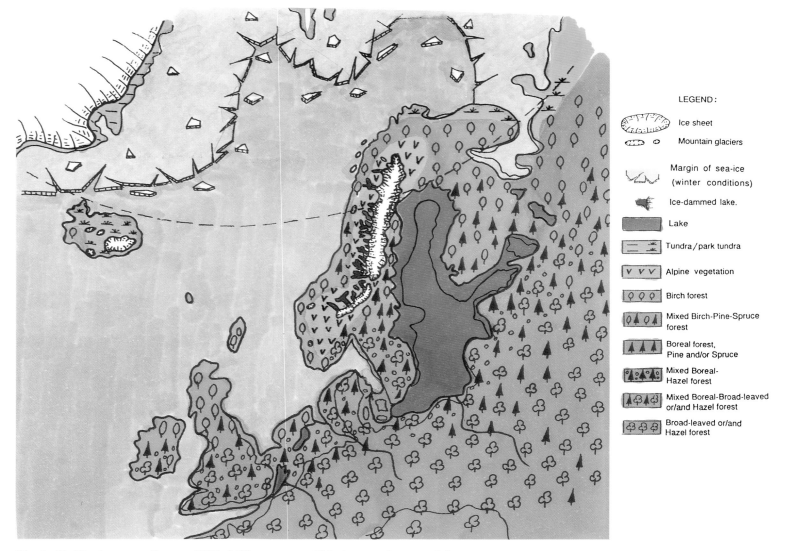

LEGEND:

Ice sheet

Mountain glaciers

Margin of sea-ice
(winter conditions)

Ice-dammed lake.

Lake

Tundra / park tundra

Alpine vegetation

Birch forest

Mixed Birch-Pine-Spruce
forest

Boreal forest,
Pine and/or Spruce

Mixed Boreal-
Hazel forest

Mixed Boreal-Broad-leaved
or/and Hazel forest

Broad-leaved or/and
Hazel forest

Fig. 2–67. Northwestern Europe, 9000–8600 years ago. This is an early part of the Ancylus Lake phase in the Baltic Sea area. The lake drained by a river through the Danish straits. Remnants of the North European Ice Sheet still remained in the mountain valleys along the Scandinavian mountain chain. The period was very warm, and the ice remnants were generally stagnant. They dammed lakes in the upper parts of the normally east-draining mountain valleys. The age of the presented lakes is probably closer to 9000 years than 8600 years. The southern part of the North Sea was still dry land, with one lake draining southwards through the English Channel Valley, and another lake draining northwards to a bay at the mouth of the extended Elbe River Valley.

Baltic Sea history (Figs. 2–66, 2–67)

The retreat of the ice front from the central Swedish end moraines, about 10 400 years ago, resulted in a rapid drainage of the Baltic Ice Lake across a low threshold that became ice free near Billingen in southern Sweden. The surface of the lake dropped about 30 m, and a broad connection with the North Sea was opened, as shown in Fig. 2–66. In this way the Baltic Sea became the marine *Yoldia Sea*, in which the marine mollusc *Portlandia (Yoldia) arctica* lived. Glaciomarine sediments of the Yoldia Sea cover much of the emerged areas of southern Sweden and Finland.

The marine connection across the central Swedish lowland gradually closed due to the isostatic uplift of the land, and the Yoldia Sea changed to a fresh-water lake, the *Ancylus Lake*, which existed from about 9000–8000 years ago. The Ancylus Lake drained through outlet rivers in the Danish straits. At about 8000 years ago the eustatic rise in world sea level, which was most likely caused mainly by the final melting of the North American Ice Sheet, resulted in a marine transgression through the Danish straits, and the Ancylus Lake changed to the salt-water *Littorina Sea*.

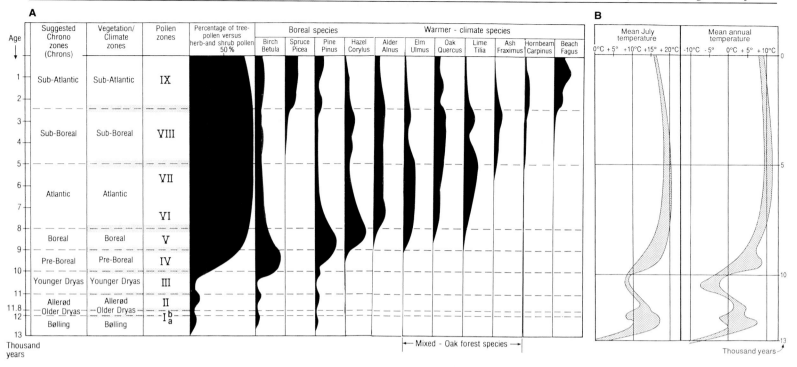

Age	Suggested Chrono zones (Chrons)	Vegetation/ Climate zones	Pollen zones	Percentage of tree-pollen versus herb-and shrub pollen 50 %	Boreal species				Warmer - climate species							
					Birch Betula	Spruce Picea	Pine Pinus	Hazel Corylus	Alder Alnus	Elm Ulmus	Oak Quercus	Lime Tilia	Ash Fraximus	Hornbeam Carpinus	Beach Fagus
1	Sub-Atlantic	Sub-Atlantic	IX												
2															
3	Sub-Boreal	Sub-Boreal	VIII												
4															
5	Atlantic	Atlantic	VII												
6															
7			VI												
8	Boreal	Boreal	V												
9	Pre-Boreal	Pre-Boreal	IV												
10	Younger Dryas	Younger Dryas	III												
11	Allerød	Allerød	II												
11.8 / 12	Older Dryas	Older Dryas	I b												
13	Bølling	Bølling	I a												

Thousand years

← Mixed - Oak forest species →

Mean July temperature: 0°C +5° +10°C +15° +20°C

Mean annual temperature: -10°C -5° 0°C +5° +10°C

Thousand years

The elevated shorelines

Some of the most striking features within the area that was deglaciated in Scotland and the Baltic Sea area and along the west coast of Sweden, the north coast of Denmark, and the entire coast of Norway are emerged shorelines. Series of shorelines were formed at gradually lower levels as the glacio-isostatic uplift progressed and time passed. All shorelines are tilted (the oldest tilt most), and they all rise towards centers of maximum uplift: one center in Scotland, and one center in the northern part of the Baltic Sea, where the highest-lying shoreline lies approximately 300 m above present sea level (see Fig. 1–15 and discussions on p. 16).

Fennoscandia becomes forested during and after the retreat of the ice sheet (Figs. 2–68, 2–69)

The Younger Dryas vegetation in the ice-free parts of Fennoscandia was mainly an Arctic tundra vegetation, with low bushes and scattered birch trees in some areas. During the following Holocene warming, trees invaded southern Fennoscandia; first came birch, followed by the Boreal forest trees dominated by pine. Later the deciduous broad-leaf/mixed

Fig. 2–68. A: Generalized pollen diagram for the southern Scandinavian region, combined with the traditional pollen zones, vegetation/climate zones and suggested chronozones. (Modified from several sources.)
Black: percentage tree pollen.
Note that a small percentage of tree pollen does not necessarily mean that the tree grew in the area, since some pollen grains can be transported long distances and some grains can be redeposited from older sediments.
All zone boundaries are time-transgressive (as indicated), except the boundaries of the chronozones. The proposed chronozones are much used in Scandinavia, but some scientists have objected to this use. If the presented names for the chronozones are used, they should be clearly indicated, such as: Boreal chronozone, Younger Dryas chronozone. (Modified from various sources.)
B: The graphs (belts) show the general trend of climate fluctuations in a region comprising Britain, Holland, northern Germany and southern Scandinavia. The graphs presented in scientific publications generally lie within the belts. They are based on the analyses of changes in flora and fauna (beetles), and the changes in distribution of various frost features.
C: Dryas octopetala, a commonly occurring plant in Arctic plant communities of late Cenozoic age. The Younger Dryas, Older Dryas and Oldest Dryas vegetation zones were named after this plant. (Photo by Leif Ryvarden.)

oak forest arrived and occupied much of southern Scandinavia during the Climatic Optimum phase of 8000 to 3000 years ago. The climatic cooling that took place after 3000 years ago led to a retreat of the mixed oak forest, and the Boreal/Subarctic spruce forest gradually expanded after the start of the cool Subatlantic period, about 2500 years ago. This expansion continued, and it is still going on today. Figure 2–68 is a typical pollen diagram which shows how the forest vegetation in an area in southern Scandinavia changed with time during the Holocene.

Figure 2–69 presents a generalized recon-

LEGEND:

- Ice sheet
- Mountain glaciers
- Margin of sea-ice (winter conditions)
- Open ocean water
- — '' — Subtropical
- Tundra / park tundra
- Alpine vegetation
- Subalpine vegetation
- Steppe/Parkland
- Parkland (partly forested)
- Desert
- Birch forest
- Mixed Birch-Pine-Spruce forest
- Boreal forest, Pine and/or Spruce
- Mixed Boreal-Broad-leaved or/and Hazel forest
- Broad-leaved or/and Hazel forest
- Mixed Oak-Pine forest on the Iberian Peninsula
- Mediterranean type forest

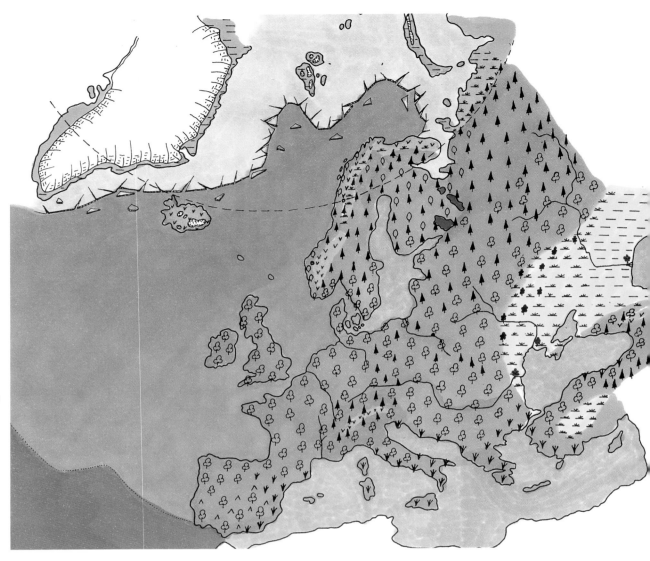

Fig. 2–69. Europe, about 4000 years ago. The presented vegetation pattern approximates the one which would prevail today if human activity had not influenced, and to some degree changed, the pattern. The ice cover on the northern part of the Atlantic Ocean represents winter-ice conditions.

struction of conditions in Europe about 4000 years ago. This reconstruction could, with minor changes, represent most of the period from 8000 years ago to the present. The changes caused by the cooling of the climate after about 3000 years ago and changes caused by human activity during the last centuries would give a somewhat different reconstruction for present-day conditions.

North America after 10 000 years ago

A marked warming took place after 10 000 years ago, and the ice sheet began to melt rapidly. However, a considerable mass of the ice sheet still remained, and it readvanced

to deposit the Cochrane end moraines of central Canada as late as about 8200 years ago. Shortly after this, the last remains of the ice sheet rapidly melted. Drumlin fields and extensive systems of esker ridges are prominent features on the Canadian Shield from this late deglaciation period, and their distribution demonstrates that the ice sheet had two domes, the Keewatin and the Labradorian (Figs. 2–70, 2–71). Ice-dammed lakes were also formed along parts of the retreating ice margin, in particular to the south and west of Hudson Bay (see Fig. 2–57).

The elevated marine shorelines

The glaciated parts of North America were isostatically depressed in the same manner as in

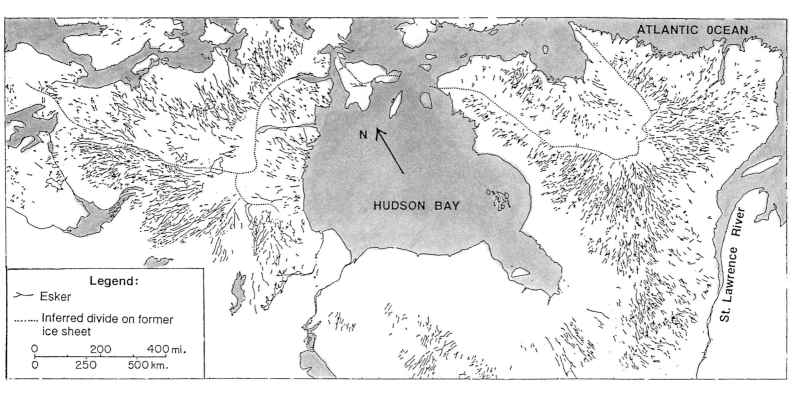

Fennoscandia, and raised shorelines exist along the glaciated coastal regions. They are all tilted and rise towards a center over the Hudson Bay region. Some of the most distinctive shorelines lie in this region, up to about 300 m above sea level (see Fig. 1–16). Even the glacial lake shorelines in the Great Lakes region are tilted and rise in the direction of Hudson Bay, which was presumably the area of thickest ice.

The vegetation

The warming of the climate after 10 000 years ago resulted in a drastic change in the vegetation pattern. The open tundra/prairie zone disappeared, and the deglaciated areas next to the ice sheet were covered by birch forest in the east, a mixed coniferous-deciduous forest in the midwest, and a wide open prairie zone further to the west. In addition the deciduous forest expanded drastically in areas to the south and southeast of the Great Lakes. As the ice sheet melted and the ice front retreated northwards, the birch forest expanded to cover most areas near the ice sheet. However, between 7000 years ago and the present, the spruce forest gradually expanded to occupy the northern forest zone; it replaced the northern part of the birch zone in most parts of

the continent, and the present-day vegetation pattern was gradually established. The warmest Holocene time is not clearly recognizable in the pollen record. In Europe this period (8000–4000 years ago) is much more distinct.

More than 15 of the larger mammal species, including the wooly mammoth, Colombian mammoth, mastodon, a camel, a tapir, and two species of bison suffered extinction about 10 000 years ago. Clearly they were hunted by paleo-Indians, and it has been suggested that this hunting was the principal cause for the extinction. However, the rapidly changing

Fig. 2–70. The distribution of esker ridges in Canada. Note how the radiation of the ridges from the dotted lines defines the position of the former ice divides. (Modified from R.F. Flint, 1971.)

Fig. 2–71. The Strange Lake Esker in Canadian Labrador. Observe also the numerous small lakes. Many of them are kettle lakes. (Photo by E. Evenson.)

Fig. 2–72. *Suggested dispersal pattern for early humans. Homo habilis (1) probably evolved in a narrow zone, mainly in the Rift Valley in Africa. Homo erectus (2) spread to a much larger area in Europe and Asia, and the Neanderthals (Homo sapiens neanderthalensis) (3) spread over a still larger area. Modern humans (Homo sapiens) (4) probably migrated as late as 14 000–13 000 years ago to northern Europe and across the Bering Land Bridge into North America. There is still much discussion about some of these migrations. (Modified from various sources.)*

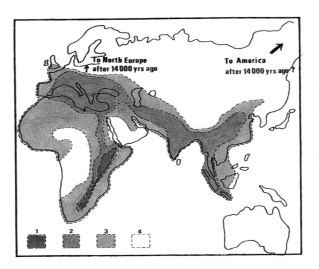

environment at that time including the expansion of forest could, at least in part, have been the reason for the extinction.

The Little Ice Age in the Scandinavian mountains

The final melting of the Scandinavian Ice Sheet took place about 8500 years ago, and even the Alpine glaciers in the high mountains probably disappeared about that time. However, evidence has been presented in favor of a regeneration of some Alpine glaciers during the moist and warm Atlantic period, between 8000 and 5000 years ago. In addition, the cooling that started 3000–2500 years ago, near the beginning of the Subatlantic period, led to a marked

readvance of the Alpine glaciers in Scandinavia, and from then on until today, Alpine glaciers have existed, and have fluctuated. In Norway the best known of these glacier readvances occurred between A.D. 1700 and A.D. 1750, during which several farms were destroyed. Most of the older Subatlantic and Atlantic marginal moraines were also destroyed during this advance. Therefore, the best record of marginal moraines is generally the one between A.D. 1700 and the present. The period between A.D. 1500 and A.D. 1930 was fairly cold with much glacier activity, and this period is frequently called the "Little Ice Age". However, the cooling which resulted in significant glacier expansion started about A.D. 1300 in many areas, terminating the late Medieval warm period around A.D. 1200. Therefore, it has been claimed that the Little Ice Age really started about A.D. 1300. However, the term Little Ice Age has occasionally been used in a wider sense for the entire cool period of the last 3000 years.

Records of Little Ice Age moraines are known from many other Alpine areas, such as the Alps and the New Zealand Alps (see p. 101). They are in general fairly similar to the Scandinavian record.

Early humans

The history of early humans is a most exciting part of the late Cenozoic history. Remains of the earlier humans such as bones and stone implements lie in cultural beds within the late Cenozoic stratigraphic system. Therefore, in most countries archaeologists work in close cooperation with other Quaternary scientists to determine the age and the environment in which early humans evolved and lived. This cooperation has been very productive, particularly in countries adjacent to the Mediterranean, where many of the significant and interesting old cultural beds are found.

The oldest fossil bones which have been classified as human come from beds in the Olduvai Gorge in the Rift Valley of Tanganyika in East Africa (Fig. 2–72). Primitive stone implements have been identified in association with the Olduvai bones. K/Ar dates indicate an age slightly more than 1.9 million years for

Fig. 2–73. *Early humans, prehistoric cultures, and tool industries. The ages presented in scientific publications for early humans and the cultures vary somewhat. Only some of the best-known tool industries are listed, and only a few characteristic tools are shown. They change from very primitive (oldest) to more refined tools.*

EARLY HUMANS	Pre-historic cultures	Age in years	Tool industries	Tools
Homo sapiens sapiens (Cro Magnon) 40.000 – 0 yrs. ago	Neolithicum	3.500		
	Mesolithicum	10.000	Magdalenien	
	Upper (Younger) Paleolithicum	40.000	Solutrien Aurignacien	
Homo sapiens neanderthalensis 125.000 – 40.000 yrs. ago	Middle Paleolithicum		Mousterien	
		125.000 ?		
Homo erectus 1,6 mill. – 75.000 yrs. ago (1,5 mill. – 0,5 mill. yrs. ago)	Lower (Older) Paleolithicum		Levalloisien Clactonien Acheulien Abbervillien	
Homo habilis 2,5 mill. – 1 mill. yrs. ago		2.5 mill.?	Olduvaien	

E

Fig. 2–74. A: The Grimaldi Museum and the Barma Grande Cave on the Italian Riviera coast. The Eemian Sea entered this cave and deposited a shore sediment with a Strombus fauna. Later, when the sea level dropped, Neanderthal man occupied the cave (see Fig. 2–75). E: Eemian marine terrace.

the Olduvai humans, who have been classified as the *Homo habilis* species. Other finds of human implements in southern Ethiopia indicate that *Homo habilis* also lived in this area as early as 2–2.5 million years ago. The later members of the *Homo erectus* species are slightly more "advanced" humans, and most finds of *Homo erectus* are also from this part of East Africa. However, fossils of *Homo erectus* have also been found in Java and in China, and were originally called, respectively, Java Man and Peking Man. If *Homo erectus* evolved in Africa they apparently migrated to the Far East and to the Mediterranean and Europe at fairly early stages. The physical character of early humans, including the shape and size of skulls, changed as time passed, and the last primitive early humans were the Neanderthals (*Homo neanderthalensis*). They lived in large areas of central Europe and in the Mediterranean region up to the Middle Weichselian time about 30 000–40 000 years ago; at that time this species appears to have disappeared rather suddenly, and the modern species, Cro-Magnon type humans (*Homo sapiens*), took over. In many cases modern humans occupied the same sites as the Neanderthals, and some scientists believe that they killed the Neanderthals, who probably were competitors for the hunting grounds. However, other scientists believe that there were other reasons for the

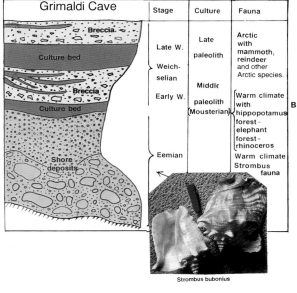

Strombus bubonius

B: The stratigraphy observed in layers below the floor of the Barma Grande Cave at Grimaldi. A warm-climate Eemian fauna was found in the lowermost part of the Mousterian culture beds, and an Arctic fauna lay in the younger culture beds. The caves were inhabited by Neanderthals in Mousterian time, and by Cro-Magnons in Late Paleolithic time. The marine Strombus fauna (C) in the shore deposits is characteristic for the Eemian Interglacial in the Mediterranean area. (Modified from a poster exhibited at the Grimaldi Museum.)

disappearance of the Neanderthals. Early humans seem to have migrated across the Bering Land Bridge to America, probably as late as 14 000–12 000 years ago.

Early humans were hunters/gatherers, and the tools which were used were mainly primitive stone implements, which changed with improved skill and needs. Figure 2–73 gives an impression of this change and the names used for the different tool industries. Early humans occupied caves, and much of our information about them comes from remains preserved in

A

B

Fig. 2–75. Neanderthals lived in and near the Grimaldi Caves and other caves in the Mediterranean area, central Europe, southern France, and northern Spain during the Mousterian period. They disappeared about 40 000–30 000 years ago and were replaced by modern humans, Cro-Magnons. (Painted by Zdenek Burian.)

cave deposits. They probably lived in open sites also, but evidence of these sites has in general been destroyed. There are many interesting cave sites, and the Grimaldi Caves on the Italian Riviera are used as an example. There the oldest cultural beds represent the Neanderthal culture, while skeletons and artifacts of modern humans are present in the younger beds (see Figs. 2–74, 2–75). Stone implements and bones of animals were found in most parts of the cultural beds. Of particular interest is the occurrence of the remains of an Arctic fauna with wooly mammoth and reindeer in culture beds from the last glaciation.

Excellent cave paintings (Fig. 2–76), especially in caves in southern France and northern Spain (Pyrenees Mountains), were painted by the Cro-Magnon humans sometime between 20 000 and 10 000 years ago, and depict early humans hunting mammoth, reindeer, and many other species. Recently, cave paintings about 27 000–18 500 years old have been discovered in a submarine cave on the French Mediterranean coast, about 36 m below present sea level. This observation corresponds well with the fact that a wide zone of the Mediterranean was dry land during the Weichselian Glaciation, and early humans lived also in this zone (see Fig. 2–19).

As the climate improved and the vegetation zones and the big game animals migrated northwards, mainly after 14 000 years ago, early humans migrated also, and cultural beds from this period have been found at several localities in northern Europe. Well known is the Hamburg culture in northern Germany, from between 12 000 and 13 000 years old. For people of this culture the hunting of reindeer

Fig. 2–76. Artistic cave paintings made by Cro-Magnons sometime between 20 000 and 14 000 years ago are found in many caves in southern France and northern Spain. A: Mammoth and goat in the Rouffignac Cave in southern France. B: Wild horses and wild oxen (cows) in the Lascaux Cave in southern France. (Photos by Jean-Pierre Bouchard.)

was very important, and the location of many early human habitation sites in northern Europe was probably to some extent determined by the migration routes of the reindeer.

Africa

The large African continent is composed of several different vegetation zones – desert, savannah and forest regions – and some of its highest mountains are glaciated today (Fig. 2–77). The late Cenozoic climatic fluctuations which resulted in glacials and interglacials in the mid-latitudes resulted in wet pluvials and dry interpluvials in the African desert regions. However, the pluvials were not necessarily in phase with the glacials, and pluvials in the various desert regions of Africa were not always in phase with each other. In addition the forest, savanna and desert regions, for example, expanded and contracted with the changing climate, as did the glaciers on the highest mountains.

Unfortunately the late Cenozoic history of many areas of Africa is rather obscure, and the observations presented are frequently conflicting. However, it is far beyond the scope of this book to reconstruct the various histories. Only a few aspects and regions will be presented in the following sections as examples.

The Sahara Desert

The Sahara is by far the largest present desert in Africa. However, much of it was covered by savanna and even some lakes during the pluvials. Cave paintings, together with stone hunting implements such as stone projectile points and scrapers which have been found over vast parts of present desert areas, show that pluvial-age humans hunted big game on the savanna which existed there at that time.

The Sahara pluvials, however, did not correspond directly with European glacials. In fact the desert region was very extensive during the Late Weichselian Glacial from 21 000 to 12 000/10 000 years ago, and the last pluvial corresponds mainly with the Early Holocene, from 10 000–4000 years ago, although it started about 12 000 years ago in some areas. Figure

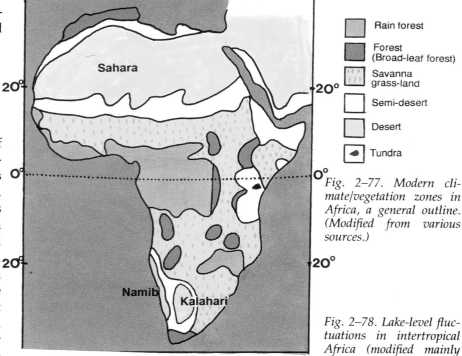

Fig. 2–77. Modern climate/vegetation zones in Africa, a general outline. (Modified from various sources.)

2–78 presents a review of pluvials and interpluvials in intertropical Africa, including the Sahara Desert.

The Namib-Kalahari region in South Africa is the second largest desert region in Africa. The central Namib Desert was arid between 20 000 and 10 000 years ago. However, lake phases seem to have existed in parts of the Kalahari Desert 30 000–25 000, 20 000–19 000 and about 12 000 years ago, and much of the time between 20 000 and 10 000 years ago was probably relatively moist.

Forests, climate and glaciations

The temperature 20 000–18 000 years ago was lower than today in most of Africa, in the order of 7°C colder in parts of eastern Africa, and the winds were stronger and generally drier. The lowland forests diminished drastically, the forest belts on the mountains shifted about 1000 m downwards, and the glaciers on the highest mountains were considerably larger than today.

Evidence of four late Cenozoic glacial episodes have been reported from Mt. Kilimanjaro. The Late Weichselian glaciers were smaller than some earlier ones, but still they covered

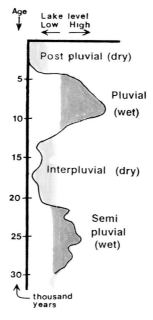

Fig. 2–78. Lake-level fluctuations in intertropical Africa (modified mainly from F.A. Street and A.T. Grove, 1979). Observe that the interpluvial (dry) phase corresponds almost perfectly with the Late Weichselian/Wisconsin glacial stage, and the pluvial (wet) phase corresponds mainly with the early, postglacial warm phase. Note also that the post-pluvial, dry phase, which includes the present day, is extremely dry.

Fig. 2–79. The Wisconsin/Weichselian maximum glaciers in South America (black). White: The largest present-day glaciers. Blue: The approximate maximum glacier extent during earlier late Cenozoic glaciations. (Modified from R.F. Flint, 1971, J. Hollins and D. Schilling, 1981, and other sources.)

Fig. 2–80. The Wisconsin/Weichselian maximum glaciers in New Zealand (black), and the land bridge which connected the South Island with the North Island. White: The present-day glaciers. (Modified from various sources.)

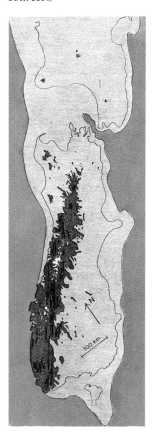

areas very much larger than those of the existing glaciers, and some glaciers descended 1000–1500 m below the level of recent glaciers. The dating of the Late Weichselian glacial episode is poor, but suggests that the maximum was probably reached about 15 000 years ago.

South America (Fig. 2–79)

South America is a large continent which is composed of desert regions, tropical rain forest, steppe, and high mountains, of which the highest are presently glaciated. The climate, vegetation zones, and the glaciers fluctuated throughout the late Cenozoic, but the information about their fluctuations is sparse.

In the Chilean-Argentinian area glacial deposits are interbedded with volcanic deposits which have been dated by the K/Ar method, yielding ages of about 5–6 million years and 3.6 million years for the oldest glacial deposits. The most extensive glaciation is more

than 170 000 years old, and at that time the glaciers were distinctly larger than the glaciers of the last glaciation, shown in Fig. 2–79. The Late Weichselian/Wisconsin (W/W) glaciers probably reached a maximum at about 20 000 years ago.

Glacier retreat probably started after 18 000 years ago, and it was followed by a readvance about 14 000–15 000 years ago. The glaciers at lower latitudes, in Ecuador, Peru and Bolivia, seem to have also reached a W/W maximum during this late advance. A considerable climatic warming and glacier retreat started about 14 000 years ago, and by about 11 000 years ago many glaciers were probably smaller than today. However, young Late Weichselian marginal moraines have been reported from parts of Patagonia, and some of them could be of Younger Dryas age.

Several lakes that lie in the western and eastern foothills of the Andes Mountains are dammed by the W/W moraines, and much of the flat Argentine pampas to the east of the mountains represent outwash plains more or less covered with eolian sand and loess blankets deposited during the various glaciations.

New Zealand

The highest mountains of the Southern Alps on the South Island are glaciated today (see Fig. 2–80). However, much larger glaciers expanded within and beyond the mountains during the last Cenozoic glaciations. Very little is known of the ages of earlier glaciations, but studies of tectonically raised marine deposits suggest that a severe cooling occurred about 5–6 million years ago, followed by a considerably warmer early Pliocene period. Severe cold periods also occurred in middle and late Pliocene and in early Pleistocene. The oldest recorded glacial deposits in New Zealand define the Ross Glaciation, which is older than 1.5–1.0 million years, while the pollen record indicates several cold phases centered on 1.3–1.0 million years ago.

The W/W glaciers reached their maximum extension between 20 000 and 18 000 years ago (see Fig. 2–80). A readvance occurred about 14 000 years ago, and at least some glaciers seem to have had a readvance just after 11 000 years ago. However, the pollen record presents no clear evidence of a Younger Dryas age cooling. A general warming started shortly after 14 000 years ago and continued into the Holocene.

Small glacial readvances occurred at some glaciers about 2300 and 1300 years ago. Marginal moraines near the existing glaciers were deposited during Little Ice Age readvances, and moraines in front of Franz Josef Glacier are dated at about A.D. 1500, 1600, 1750, 1820, 1880, 1909, 1934, 1954, 1968 and 1989.

Many large lakes in the eastern foothills of the Southern Alps on the South Island are dammed by the W/W moraines, and the plains which apron much of the eastern part of the island to the east of the lakes are outwash plains deposited by the glacial rivers.

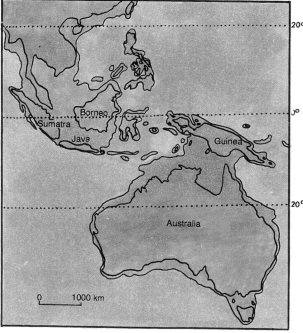

Fig. 2–81. Indonesia and Australia during the Late Weichselian maximum. Land bridges connected several of the large Indonesian islands with the mainland, and Australia with New Guinea and Tasmania. (Modified from various sources.)

Australia, New Guinea and Tasmania

The late Cenozoic climatic fluctuations certainly affected Australia and adjacent islands. Lacustrine sediments observed in the large central Australian desert are known to represent pluvial phases, but the phases have not been well dated.

Small cirque glaciers formed in the highest Australian coastal mountains, while larger glaciers occurred in the mountains in New Guinea and Tasmania during the W/W Glaciation. Radiocarbon dates indicate that these glaciers reached a maximum extent about 20 000–18 000 years ago in Australia and Tasmania, and sometime before 15 000 years ago in New Guinea, followed by two readvances which culminated shortly after 13 000 and 11 400 years ago. Tasmanian glaciers were much larger during an earlier glaciation.

During the global sea-level lowering associated with the W/W Glaciation, a land bridge existed between Australia and New Guinea, which allowed animals to migrate, accounting for much of the exotic endemic fauna found in both Australia and New Guinea (see Fig. 2–81).

Shorelines about 4 m above present sea level along the Australian coast represent the higher sea-level stand during the E/S Interglacial, about 125 000 years ago. Of particular interest

are the tectonically raised marine terraces and coral reefs on New Guinea. The reefs/terraces have been U-Th dated, and on the basis of the dates, the measured altitudes, and a calculated rate of tectonic uplift, a graph for sea-level fluctuations was established (see Fig. 3–53). The graph shows that the sea level had a maximum low, about 100–120 m below present level, about 18 000 years ago.

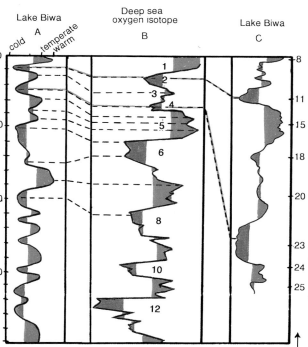

Fig. 2–82. Climate fluctuations recorded in sediment cores from Lake Biwa, Japan. Two graphs of climate fluctuations at Lake Biwa correlated with a deep-sea oxygen-isotope record and isotope stages. The Biwa graphs are based on biostratigraphy, mainly pollen stratigraphy in a 200 m long core (A) and a 25 m long core (C). The deep-sea graph (B) is modified from N. Shackleton and N. Opdyke, 1976. The Biwa graphs are modified from N. Fuji, 1974. Note the good correlation with the isotope graph. Blue: cold, or relatively cold. Red: warm, or relatively warm.

Fig. 2–83. A: Weichsel-ian/Wisconsin glaciers in Asia. This map recon-struction was made some years ago for the CLIMAP project, and it corresponds approximately with recon-structions which were accepted by many scien-tists in the former Soviet Union.

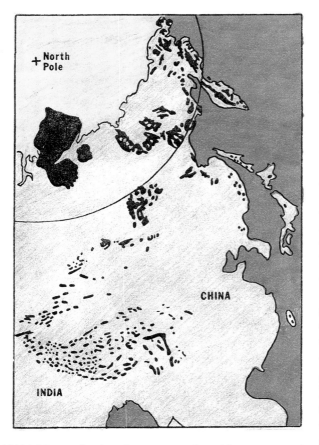

B: The suggested maxi-mum extent of the Late Weichselian ice sheets in Siberia (according to M. Grosswald, 1988) and on the Tibetan plateau (according to M. Kuhle, 1988). White: ice sheets. Violet: ice-dammed lakes. Dark green: other lakes. Pale blue: ice shelf. K: Kara Sea Ice Sheet. E: East Siberian Ice Sheet. T: Tibetan Ice Sheet. (Modified from M. Gross-
wald, 1988, and M. Kuhle, 1988.) Many scientists do not accept the evidence presented in favor of the large Late Weichselian ice sheets. They claim that most field evidence is in disfavor of this theory. However, the theory is fascinating, and it is now being tested.

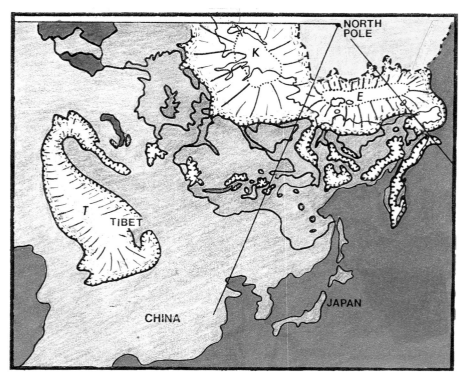

Antarctica

The Antarctic Ice Sheet fluctuated essentially in phase with the mid-latitude northern hemi-sphere ice sheets. In particular the marine-based sections of the ice sheets and associated ice shelves expanded substantially, at least during the two last main northern hemi-sphere glacials. During the W/W Glacial a maximum was reached sometime between 22 000 and 17 000 years ago. However, the small local glaciers and other glaciers which terminated on land were, in general, smaller during the glacials than today. Even parts of the large East-Antarctic Ice Sheet were proba-bly slightly thinner, or about as thick as today, during the glacials.

In the cold climate of Antarctica the surface snow-melting is, in general, of minor impor-tance to the mass balance of the ice sheets, while snow accumulation and calving are the two main factors which determine the ice-sheet budget. This explains the different be-havior of different sections of the ice sheet and of the local mountain glaciers. The sizes of the calving glaciers are largely governed by sea-level fluctuations, and they expanded during the low sea-level periods of the glacials and retreated during the interglacials. How-ever, the climate was colder and the snow accumulation was in general less during the glacials. Therefore, the glaciers which were not, or very little, influenced by calving usually retreated during the glacials.

Analyses of the ice core, which was collected at the Vostok Station on the central polar pla-teau on the East-Antarctic Ice Sheet, indicate that the air temperature has fluctuated very much in phase with the general worldwide trend of climate fluctuations (see Fig. 1–24). It was much colder around 18 000 years ago than today, while the precipitation at that time was considerably less than today. See also p. 25.

Asia

The Asiatic region is so vast in area and it is composed of so many different climate regions and vegetation zones that it is impossible to present a realistic synthesis within the scope of this book. In addition, the results of the re-search on the late Cenozoic climate and glacier fluctuations is still fragmentary, at best, from most regions. Therefore, only a few main trends and generally accepted records will be mentioned.

Available records show that the climate, vegetation zones, and glaciers within the entire region fluctuated more or less synchronously with the general worldwide climate changes recorded in oxygen-isotope and pollen diagrams, of which many have been presented in previous sections. As the climate became colder, the cold-climate vegetation zones and glaciers expanded during the glacial periods, and they decreased when the climate became milder during the interglacials.

Of particular interest is the newly observed stratigraphy of the loess plateau in central China, where a complete succession of late Cenozoic glacial loess beds and interglacial soils has been recorded (see p. 22).

Another important late Cenozoic stratigraphic succession has been recorded in more than 200 m long cores from Lake Biwa in Japan. The graphs in Fig. 2–82 record the temperature fluctuations through several hundred thousand years, based on pollen studies and various other kinds of analyses. Note how well the graphs correspond with, for instance, the oxygen-isotope graph.

In the mountainous Verkhoyansk region and the Yenisei region of Siberia, relatively warm-climate Middle Weichselian interstadial deposits have been recorded, and even deposits corresponding with the Younger Dryas cold phase and the Bölling and Alleröd warm phases are present. In addition, it has been indicated that Early Weichselian glaciers (isotope stage 4?) and Middle Weichselian glaciers were larger than the Late Weichselian glaciers. However, there has been much discussion and controversy about the size and timing of the Asian glaciers and glacier events. Figure 2–83A presents a reconstruction of the Asian glaciers during the Late Weichselian maximum. This reconstruction corresponds approximately with the view held by most Russian scientists. Recently some scientists suggested that more-extensive Late Weichselian ice sheets covered parts of Siberia and the Tibetan Plateau. However, their reconstructions, which are outlined on Fig. 2–83B, have been questioned by many scientists, who claim to have field observations which exclude the possibility of large Late

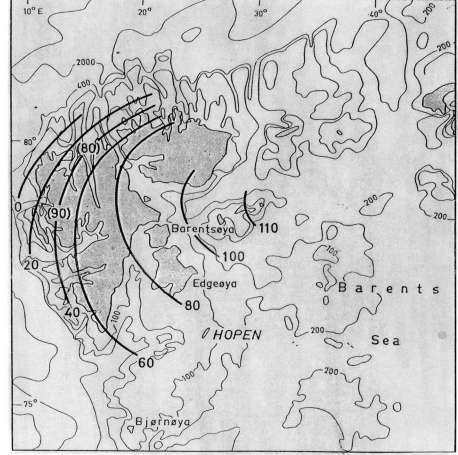

Fig. 2–84. Isolines showing the postglacial isostatic uplift (in meters) in the Spitsbergen area. The lines for 0 m, 20 m and 40 m represent about 13 000–11 000 year old shore levels, and the lines for 100 m and 110 m represent about 10 000 year old levels, since this area was deglaciated that late. Note that the amount of uplift increases towards the Barents Sea area, suggesting that this area was the most isostatically depressed by the ice sheet during the Late Weichselian. (Modified from various sources.)

Weichselian ice sheets. Several of them accept the evidence for a former extensive Kara Sea Ice Sheet, but most observations seem to indicate that this is of pre-Late Weichselian age. The existence of large East Siberian and Tibetan ice sheets of Late Weichselian age seems even more questionable, according to most scientists.

The Arctic region

Greenland, Spitsbergen and the Barents Sea

Some of the best records of Weichselian/Wisconsin climate flucutations were obtained from glacier ice cores from the ice sheets on Greenland and Baffin Island. They show a remarkably good correlation with the climate fluctuations recorded elsewhere (see p. 23). However, the records of glacier-margin fluctuations from these areas are much more fragmentary and, in some cases, still confusing. The glaciers in the Canadian Arctic, in Greenland, and in Spitsbergen were considerably larger than today during the W/W Glaciation. However, there has been discussion about how much larger they were during their maximum extension, and when that maximum occurred. The oldest distinctive end moraines on land in most regions seem to be somewhere between 12 000 and 9000 years old. They are located from the middle to near the mouths of the fjords, and this shows that considerable parts of the coastal land areas were ice free at that time. The presence on these coasts of emerged beach ridges and marine deposits more than 25 000 years old, which appear not to be glacially tectonized nor covered with till, has led many scientists to conclude that the oldest distinc-

tive end moraines on land indeed represent the maximum extension of the Late W/W glaciers. However, other scientists have now claimed that the Late W/W glaciers could have covered much of the coastal area and the mentioned marine features without leaving much evidence. If the glaciers were cold-based and frozen to the ground, they could have expanded over the coast without tectonizing the shore features nor depositing much basal till. In addition, recent studies of marine deposits related to rather well-developed submarine end moraines on the shelf off the coasts suggest that they are of Late W/W age.

Recent studies of sediment cores from the Barents Sea indicate that it was covered with a grounded ice sheet, and several radiocarbon dates of deposits connected with high-lying isostatically raised shorelines (Fig. 2–84) on the islands to the east of Spitsbergen show that a considerable ice dome must have remained over the northern part of the Barents Sea, including those islands, as late as 11 000–10 000 years ago.

Chapter 3
PROCESSES AND SCIENTIFIC METHODS

The history presented in Chapter 2 is based on information obtained through elaborate field and laboratory studies, including studies of modern processes. Chapter 3 is focussed on some of the most important processes and methodologies used to reveal this history.

PROCESSES

Glaciers form landscapes; the glacial environment

Much of the late Cenozoic ice age history in Europe and North America relates to glacier distribution, chronology and activity. Some glaciers are able to erode, transport and deposit large amounts of material at a speed which is unequalled by most other geological processes. Combined with the increased erosion by frost, wind and glacial rivers in a glacial environment, many late Cenozoic-age glaciers were able to form and transform landscapes within very short periods of time, geologically speaking. To understand how this could happen, a basic understanding is needed of glacial and periglacial processes. The following is a brief introduction to these subjects, with most attention paid to the geologic results of the processes. More detailed and more in-depth discussions of current glacial and periglacial processes are presented in, for example, several textbooks listed on pp. 195–96 .

The formation of glaciers

The formation of glaciers depends upon climate, altitude, and exposure of the terrain surface. Accumulation of snow, mostly during the

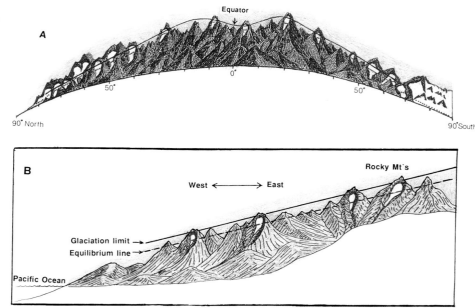

Fig. 3–1. A: *The trend of the glaciation limit (thick line) and the equilibrium line (dotted lines) from the North Pole (90°N) via the equator (0°) to the South Pole (90°S), observed along a mountain range, such as the one along the west side of North or South America.*

B: *A cross section of the glaciation limit (thick line) and the equilibrium line (dashed line) from a coastal area (here, the Pacific coast) towards a continental area (here, the Rocky Mountains). Note that the equilibrium line lies below the glaciation limit (glaciation threshold).*

winter season, and the melting caused by the summer heat are two main factors that determine the presence and health of glaciers. High snow accumulation combined with low summer heat and melting promotes glacier formation and expansion. In glacial geology the term *"glaciation limit"* or *"glaciation threshold"* represents an imaginary surface above which there is perennial snow and glaciers, and whose altitude depends upon the two mentioned climatic factors. During a climatic

Fig. 3–2. The equilibrium line (the dotted line) on a glacier near Zermatt in Switzerland. There is a net annual accumulation of white snow above the line, and a net annual ablation below the line, where darker firn and old ice occur on the surface. The accumulated snow is gradually transformed to firn and ice as it passes through the glacier. (Photo by Nils Haakensen.)

cooling this surface drops, and when it reaches the altitude of the highest mountain summits, small glaciers may form on these mountains (see Fig. 1–30 and the Glossary). With a further drop glaciers form at successively lower altitudes, and they may merge to form larger ice caps or sheets. This happened in the mid-latitude mountains of the northern hemisphere during the start of each late Cenozoic glaciation. Figure 3–1 shows how the altitude of present-day glaciation limits, and snow lines, rises from the polar regions towards the low-latitude dry desert regions and lowers some-

what again in the wet tropical regions. In addition the altitude rises from the moist coastal regions towards the dry continental regions, as, for instance, from the west coast of North America to the drier eastern parts of the Rocky Mountains. As seen in Fig. 3–1 the glaciation limit lies about 100–300 m above the snow line or equilibrium line, which is described in the following section.

The glacier budget

All glaciers consist of snow, firn, and ice with an admixture of rock debris, and all particles in an active glacier are transported through the glacier along "imaginary" flow lines (see Fig. 3–3).

The glacier budget represents the balance between the gain of snow, firn, and ice and the loss of snow, firn, and ice from the glacier. Depending upon the amount gained (accumulated) and lost (ablated), the glacier can have a positive, a balanced, or a negative budget at any given time.

Snow which accumulates on the glacier surface, mainly during the winter season, partly melts during the summer season. However, on the higher part of the glacier surface, there is an annual layer of snow that does not melt completely, and this layer, together with ice formed by freezing of meltwater within the glacier, generally represents the net annual accumulation. On the lower part of the glacier surface, the snow accumulated during the winter melts during the melting season, together with a lay-

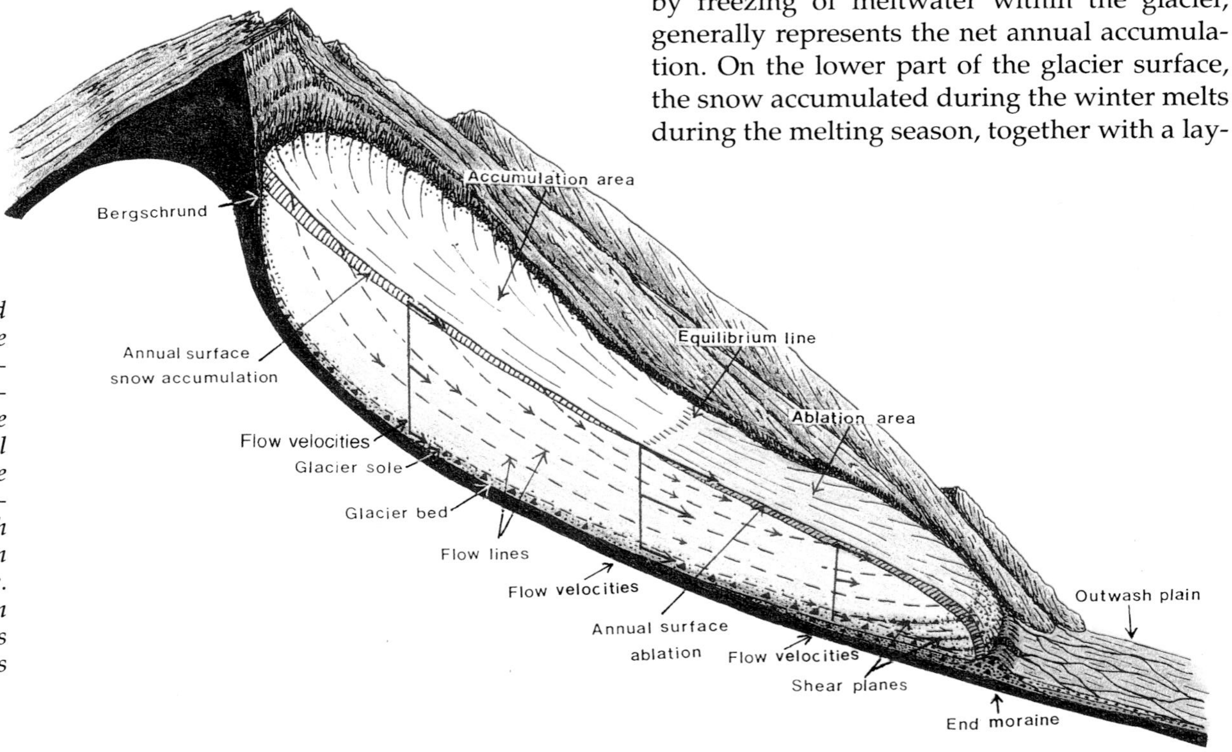

Fig. 3–3. Features related to a cirque glacier. The same features relate to valley glaciers. Flow velocities (thick arrows) are shown in three vertical sections. The arrows at the base of the glacier represent the basal slip, which usually has a maximum at the equilibrium line. Therefore, the maximum erosion generally takes place in a zone near this line.

Accumulation area

Bergschrund

Equilibrium line

Annual surface snow accumulation

Ablation area

Flow velocities

Glacier sole

Glacier bed

Flow lines

Flow velocities

Annual surface ablation

Flow velocities

Outwash plain

Shear planes

End moraine

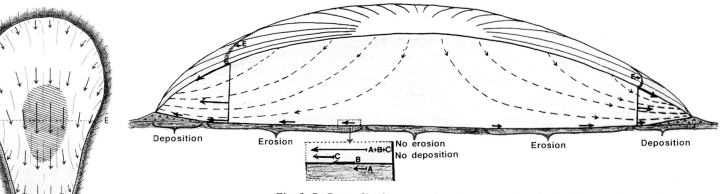

Fig. 3–5. *Generalized cross section through an ice sheet with margins resting on land. Dashed lines = flow lines. Heavy arrows = flow velocities. Heavy arrows at the base = the sum of A+B+C. A: Deforming motion within the sediments below the glacier bed. B: Basal slip (at the glacier bed). C: Creep/slip within the lowermost ice layers. E: Equilibrium line. The vertical scale is very much exaggerated.*
The zones of erosion and deposition vary in width, depending upon the temperature, shear stress and flow velocity at the base. No, or very little, erosion takes place when the ice is frozen to its bed and both A and B are zero.

Fig. 3–4. *Vertical view of the same cirque/valley glacier which is shown in Fig. 3–3. The arrows indicate the glacier-flow velocity at/near the surface of the glacier. Maximum glacier-flow velocity occurs within the shaded area.*

E-E: The equilibrium line. M: The marginal moraine.

Observe:

1. The maximum glacier-flow velocity usually occurs in the center of the glacier near the equilibrium line.

2. The flow velocity decreases towards the glacier margin.

3. The glacier flow is directed away from the margin and into the glacier above the equilibrium line, and towards the margin below this line. For this reason the marginal moraines form only below the equilibrium line.

4. The marginal moraines are very small near the equilibrium line; they are called lateral moraines between E and M, and end moraines between M and M.

er of firn and ice. This layer, in addition to ice which melts within and at the base of the glacier, generally represents the total annual ablation. However, if the glacier terminates in water (an ocean or a lake), the melting and breaking off of ice pieces (calving) from the part in contact with water represent important ablation factors. In fact, at several calving glaciers in very cold regions, this factor represents almost all of the annual ablation, since practically no significant melting occurs on the ice surface.

The line, or zone, across the glacier surface which separates the accumulation area from the ablation area, as measured at the end of the summer melt season, is called the snow line (Fig. 3–2). Unfortunately the term snow line has been used in many different ways, and therefore the term *equilibrium line* is now more commonly used for this zone.

The living glacier; glacier flow
(Figs. 3–3 to 3–6)

In the accumulation area one annual layer of snow is piled on top of another, and gradually the snow is transformed via firn to ice as the layer is buried deeper and the pressure increases. When the pressure becomes high enough, generally at a depth of 30–50 m, the ice becomes semi-viscous and starts "flowing". This "flow" occurs within the ice in the lower part of the glacier. The upper, more rigid part, near the glacier surface, rides on the moving semi-viscous ice in the lower part, and crevasses usually form only in this rigid upper part when this brittle zone is subjected to tension. The internal glacier flow prevents the formation of crevasses in the lower part.

Gravity is the ultimate driving force behind glacier flow, and the glacier generally moves in the dip-direction of the glacier surface. However, the slope of the glacier surface does not necessarily correspond with the slope of the subglacial terrain. In fact, it is one of the most typical and important qualities of glacier flow that ice in the basal parts can be driven up subglacial slopes. By this process a thick glacier, for instance a thick ice sheet, can move more or less independently of the topography of the subglacial terrain. Still, the subglacial terrain (in particular a deep valley) generally directs much of the flow in ice sheets. Observations

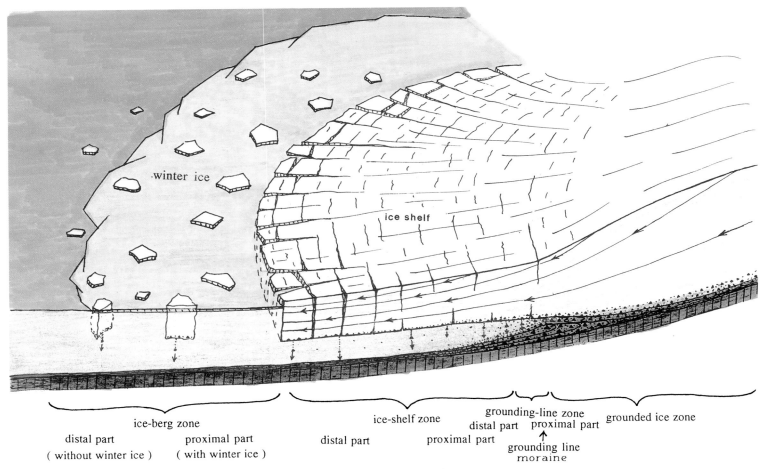

winter ice

ice shelf

ice-berg zone

distal part
(without winter ice)

proximal part
(with winter ice)

ice-shelf zone

distal part

proximal part

grounding-line zone
distal part   proximal part

grounded ice zone

grounding line
moraine

Fig. 3–6. An ice sheet which is grading into an ice shelf. This was common along the marine sectors, on the continental shelves, in both North America and northern Europe. Large ice shelves are formed only in very cold polar climates where the entire ice-shelf surface generally lies within the accumulation zone. In fact, ocean water at the base of the ice shelf may freeze to the base and add to the accumulation. Icebergs from the Antarctic ice shelves are usually fairly clean. They contain very little rock debris at the base, and accumulation of sediments in the grounding-line zone is very low.

Much of the flow of the grounded part of the glacier, near the grounding-line, is caused by shearing and deformation of the sediments at and below the glacier sole. Basal till is dragged/squeezed into the grounding-line zone where it gradually accumulates, and some of it may slide down the slope below the barely floating proximal part of the ice shelf. The maximum glacier-flow velocity occurs at and beyond the grounding-line.

from Antarctica show that, even below a fairly smooth ice-sheet surface, there can be ice streams in subglacial valleys draining much of the ice.

Resistance against the flow within the glacier ice and friction between the base of the glacier and the glacier bed retard the glacier flow. Therefore, the flow speed in general decreases from the glacier surface and downwards to the glacier bed. The shear stress increases downwards, which promotes slip between the glacier and its bed, and the glacier slides on the bed. However, some *polar glaciers*, in particular thin polar glaciers, are cold-based and frozen to the bed. At the base of this kind of glacier there is no slip and therefore little or no erosion. Beneath thicker polar glaciers, such as some of the Antarctic glaciers, the temperature at the base might still be at the pressure melting point, and the glaciers slide on a water-saturated deforming sediment bed, or slip on a bedrock surface.

The resistance against the flow within a glacier decreases with increasing ice temperatures, and therefore the "warmest" glaciers, the *temperate glaciers*, in general flow fastest. In addition, the speed of glacier flow increases with increasing dip of the glacier's top surface, increasing glacier thickness, and increasing smoothness of the glacier bed. Mathematical formulas have been determined for the speed of the glacier flow, but they are not presented here.

The observed internal flow speed of different glaciers varies considerably, from a few millimeters to several tens of meters a day. *Surging* glaciers may flow even faster, and observations in present-day glaciated regions indicate that surges are rather common. Surging appears to be caused by an "uncoupling" of the glacier from its bed by the development of a water layer. This in turn reduces the frictional drag of the ice on its bed towards zero. In general polar glaciers flow more slowly than temperate glaciers.

The imaginary flow lines in Figs. 3–3 to 3–6 show pathways along which the ice and rock debris within a glacier are transported. Note that the flow lines extend from the surface and down into the glacier within the accumulation area, and up to the ice surface in the ablation area of a glacier that terminates on land. Note also that flow lines extend from the sides and into the glacier within the accumulation area, and out towards the sides within the ablation area. The pattern of the flow lines explains several of the most important features of glacier activity, and it will be discussed in later sections.

Polar glaciers which terminate in water are frequently devoid of surface ablation, and calving is the main ablation factor. Some of these glaciers terminate at the grounding-line and some at the front of floating ice shelves. A net surface accumulation usually occurs even on the shelves, and therefore the flow lines generally extend from the surface down into the shelves (see Fig. 3–6). At many ice shelves the temperature is so low that sea water freezes to the base and thus adds to the accumulation.

The arrows in Fig. 3–4 represent the flow velocity at the surface of a glacier. Observe that the velocity is generally highest near the equilibrium line on glaciers that terminate on land, and it decreases to nearly zero near the terminus. However, on a glacier which grades into an ice shelf, the velocity increases towards the shelf and reaches a maximum on the shelf.

Geologically, the flow velocity at the base of the glacier, the basal-slip speed, is much more important than the surface velocity. As already mentioned, polar glaciers which are frozen to the bed have no basal slip, but thick polar glaciers may have a significant basal slip, and the thick, fast-flowing subpolar and temperate glaciers may have a considerable basal slip. Melted-based glaciers have in general a faster slip than frozen-based glaciers, and it has been suggested that some glaciers may have a thin film of water at the base, which promotes a fast basal slip. However, in this case the water film

may reduce the erosion rate at the base. Measurements that have been carried out on some thin temperate Alpine glaciers show that the basal-slip velocity can be as high as 80–90% of the surface velocity.

Glacier fluctuations and stagnant glaciers

The longitudinal surface profiles of land-based glaciers with balanced budgets do not change, and glaciers always strive to obtain a balanced budget. When the climate changes, and the budget becomes positive or negative, the glacier reacts with an enlargement or a reduction in the size of the ablation area, and this is achieved by an advance or a retreat of the glacier front. Therefore, a glacier is a rather sensitive climate indicator. However, the first reaction of a glacier to a changed budget is generally a rise or a drop of the glacier surface, while the advance or retreat of the glacier front is somewhat delayed. The fronts of small glaciers respond faster to climate changes than the fronts of large ones. The delay time for the response of small glaciers is generally only a few years, but for larger glaciers and ice sheets the delay can be in the order of several hundred years. If the climate warms rapidly, as it did several times during the deglaciation of the Late Weichselian/Wisconsin ice sheets, the decrease of the accumulation area and the increase of the ablation area can be so drastic that a considerable portion of the near-front (distal) part of the glacier stagnates and gradually melts as dead ice. In some cases the entire accumulation area diminished to nothing, and the entire glacier stagnated and melted as dead ice.

For glaciers which terminate in water the changes in water level can be as important as, or even much more important than, climate changes as a cause of glacial fluctuations. This is true for much of the present Antarctic Ice Sheet, and it was true for most of the marine-based parts of the late Cenozoic ice sheets in the northern hemisphere also.

Different glacier types (Figs. 3–7 to 3–11)

Glaciers can be classified according to morphology or climate/geography.

Fig. 3–7. Small hanging glaciers in New Zealand.

and ice sheets may have outlet valley and fjord glaciers. Glaciers which terminate in water may grade into floating ice shelves (see p. 108 and the Glossary). The most important glaciers for the ice age history of North America and northern Europe are the ice sheets, and they will be described in more detail later.

Glacial erosion

Glacial erosion takes place at the base of the glaciers. In addition erosion occurs at head-walls (*bergschrunds*), side-walls and nunataks, frequently in connection with frost shattering of rock along the glacier margin. *Erosion at the base of a glacier* is caused either by abrasion resulting from the friction of the glacier sole on the glacier bed, or by quarrying (plucking) of rock pieces from the glacial bed.

The *climatic/geographic* definition was described by Ahlman, who identified *polar, subpolar* and *temperate glaciers* (see Glossary).

The *morphologic definition* is based on the size, shape, confinement, and location of the glacier. The following are some main types: hanging, cirque, valley, fjord and plateau glaciers, as well as ice sheets. Valley glaciers may grade into piedmont glaciers, and plateau glaciers

The glacier sole with its numerous small and large rock fragments acts as a gigantic piece of "sandpaper" which polishes and scratches the rock bed. Evidence of this activity is the numerous polished, striated, grooved and fluted rock surfaces which are found almost everywhere within the formerly glaciated regions (see p. 15). The quarrying of rock pieces from the rock bed may result from mechanical or frictional plucking of rock pieces from the lee-slopes on the bed, but more commonly

Fig. 3–8. Plateau glaciers with outlet valley and fjord glaciers in a fjord district of eastern Greenland. Other pictures of glaciers are shown in Figs. 1–29 (cirque glacier), 2–20, 2–29 (ice sheet), 2–27, 2–28 (outlet glaciers and ice shelves), and 3–13 (valley glacier).

the plucking mechanism is ascribed to the melt/freeze process which takes place at the base of a dry/wet/freeze-based glacier where the temperature is close to the pressure melting temperature. A small increase of the basal pressure on the stoss-slope of a projecting bedrock surface may lower the melting point enough for the basal ice to melt. The water that is formed in this way on the stoss-slope may freeze when it arrives on the lee-slope, where the pressure is reduced. In this way rock pieces on the lee-slope may freeze to the base of the glacier and be pulled away. The freezing of water in the cracks on the rock bed on the lee-slope also promotes the breaking loose of rock pieces. Good evidence of the basal glacial plucking are the many characteristic steep lee-slopes on rocks in formerly glaciated regions, in particular the lee-slopes on whale-back rocks (see Fig. 1–11).

Discussion has been continuing about which is the more important, glacial abrasion or glacial plucking, in forming the large glacial landscape features, but with no conclusive result. What we observe is the result of the combined effect of the two, and that the magnitude of the erosion is impressive in many cases.

We can conclude that the amount of glacial basal erosion depends upon several factors, such as:
1. the shear stress and basal-slip velocity;
2. the pressure at the base, which depends largely upon the thickness of the glacier;
3. the character of the sole with its rock fragments (hard and large clasts promote abrasion, and soft, small clasts do the opposite);
4. the character of the bedrock or sediment at the glacier bed (sediments and soft and/or jointed rocks are less resistant);
5. the temperature at the base (a temperature near the pressure melting point promotes erosion);
6. the presence of water at the base (a film of water under high pressure reduces erosion);
7. the time involved.

Where does glacial erosion take place?

Glacial erosion does not take place everywhere under a glacier, and not even under all glaciers. As already mentioned, glaciers or parts of glaciers which are frozen to the bed may not

Fig. 3–9. The margin of the ice sheet in western Greenland. Similar conditions existed along parts of the terrestrial margins of the North American and North European ice sheets. However, considerable parts of the ice-age ice margins in Europe and North America lay on ground which was relatively flat, and flat outwash plains frequently extended beyond the ice margins (see Fig. 3–22). Note the end-moraine ridge which fringes the largest glacier lobe. (Photo by Henrik Höjmark Thomsen.)

erode, and deposition rather than erosion generally takes place in a peripheral zone near the front of most land-based glaciers. Observations from regions formerly glaciated by ice sheets show that very little or no erosion took place below parts which lay at or near the ice divide. This corresponds well with conclusions based on flow-line reconstructions (see Fig. 3–5). The main erosion of land-based glaciers usually occurs in the intermediate zone, between the central and peripheral zones, and maximum erosion generally takes place within the zone of maximum basal slip (see Figs. 3–3, 3–5). However, variations in some of the erosional factors may change this picture somewhat. As a result of the differential erosion, and the fact that maximum erosion generally occurs at some distance behind the glacier margin, the glacier is able to erode deep rock basins or troughs (see next section).

During the coldest-climate glacial phases the large ice sheets which covered North America and northern Europe probably had broad marginal zones where the bases of the glaciers were more or less frozen to the beds. Little glacial erosion took place in these zones, and no erosion occurred in the central zones near the ice divides. As a result the intermediate zones with significant glacial erosion were probably relatively narrow. However, the extensive meltwater outwash terraces that are connected with several of the largest old end-moraine

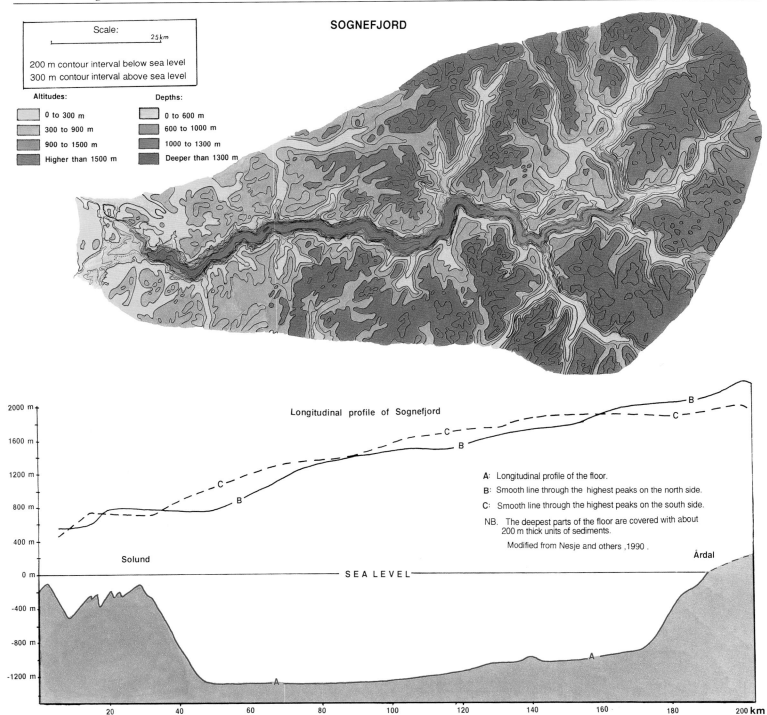

Fig. 3–10. *Sognefjord in western Norway. The rock floor is about 1500 m deep at its deepest, and the fjord sides rise to 1200–1800 m above sea level. The rock thresholds near the mouth of the fjord are 150–300 m deep. Calculations suggest that at least 2000 km³ of rock were eroded and transported out of the fjord with the glaciers and glacial rivers. This corresponds to about a 4 m thick rock layer covering most of France, for instance. Transformed to marine clay, this layer would be more than 8 m thick.*

zones in Europe and North America, and the many stream-eroded tunnel valleys in Europe, indicate that a considerable melting took place, at least during phases of the early deglaciation periods. In this kind of environment the marginal freeze zone could have been much narrower and the intermediate erosion zone much wider than indicated, even though the air temperature was low on the plains to the south of the glaciers.

Glacial-erosion features

Fjord landscapes (Fig. 3–10)

Good examples of gigantic differential glacial erosion are the many deep fjords in formerly glaciated regions. For instance, the Sognefjord in western Norway has a rock floor about 1500 m deep which is covered with a sediment unit about 200 m thick. The rock threshold at the mouth of the fjord is only 150–200 m deep, but the deepest rock passage near the mouth is about 300 m. Therefore, the rock basin behind the threshold, which must have been eroded by glaciers, is about 1200 m deep. The volume of this basin is approximately 430 km³. However, glacial erosion undoubtedly lowered the rock thresholds at the mouth of the fjord also, and much glacial erosion occurred in the part of the fjord which lies above the rock basin, where the fjord sides rise up to 1000–1500 m above sea level. Moderate estimates suggest that the total volume of glacially eroded rock in Sognefjord was more than 2000 km³, and one should keep in mind that the rocks in

Fig. 3–11. Lago Argentino occupies the deepest known probably glacially eroded basin in the world. The depth of the lake in front of Uppsala Glacier (A) is more than 1000 m. The height of the floating vertical glacier front is nearly 600–800 m, of which about 60–80 m is visible above lake level (B). Uppsala Glacier is a calving glacier, and rather large icebergs break off from the front (A). Since Uppsala is a temperate glacier, the floating part is very narrow, and no really large ice shelf exists.

Moreno Glacier (C) is another spectacular glacier which enters Lago Argentino. It has a vertical 40–60 m high front, above lake level. The front is partly grounded, and therefore mainly small ice blocks break off and float away as "icebergs". Observe the splash in the water at two places, caused by small ice blocks which dropped from the front, and note the badly crevassed surfaces of both Moreno and Uppsala glaciers. Fast-flowing glaciers which terminate in deep water are frequently crevassed like this. Their flow velocity is highest adjacent to the glacier front, and crevasses are caused by tension in the ice.

Many ice age glaciers in fjords and lakes were probably of this kind, in particular the temperate glaciers which existed during the late deglaciation phases.

Fig. 3–12. Lake landscape. Ice-sheet erosion on relatively flat bedrock terrain resulted in numerous irregular small and large shallow rock basins, as, for instance, on this mountain plateau in southwest Norway, where the bedrock consists of very hard, crystalline gneisses and granites. (Photo by Norsk Fly og Flyfoto A/S.)

Sognefjord are of hard crystalline composition. If the volume of glacially eroded rock were spread evenly over France, for example, it would cover the country with a layer at least 4 m thick. If the rock were transformed to marine clay, the layer would be more than 8 m thick.

How can glaciers form a large fjord like Sognefjord? It must have been formed during several glaciations, both by large ice streams within the ice sheet and by outlet glaciers from the ice sheet. Calculations based on the ice volumes recorded in deep-sea oxygen-isotope graphs suggest that Sognefjord could have been glaciated for more than 1 million years. Other large fjords exist which are even deeper than Sognefjord. For instance, Byrd Glacier, which is a major outlet glacier from the East-Antarctic Ice Sheet, occupies a fjord which is about 2800 m deep in central parts and only 500 m deep at the grounding-line near its mouth (see Fig. 2–27). Another good example is Northwestfjord, a branch at Scoresby Sound in eastern Greenland. The deepest part of this fjord is about 1498 m, and the mouth of Scoresby Sound is only 450–500 m deep. In addition several other fjords in Greenland are more than 1200 m deep, and several fjords in Norway are between 600 m and 1000 m deep.

Skjaergaard and fjärd landscapes (Fig. 1–32)
Coastal districts where the preglacial terrain was flat or undulating with no deep and steep valleys were frequently transformed to skjaergaard and fjärd landscapes by ice-sheet erosion during the ice ages. A skjaergaard consists of glacially sculptured skerries, islands and sounds, and fjärds are glacially sculptured "bays" with low, gently sloping sides, irregular coastlines, and floors with shallow rock basins.

Glacial-lake landscapes
Glacially eroded basins/troughs represent a most characteristic glacial-erosion feature. The troughs are generally smooth with U-shaped cross profiles. When located above sea level the basins are generally occupied by lakes, and most of the scenic lakes in the world are glacially sculptured. Ice sheets which cover flat or gently undulating terrain generally erode wide and shallow basins with irregular contours. Today such basins, filled with lakes, represent the glacial-lake landscapes which cover large parts of, for instance, Canada and Fennoscandia (see Figs. 3–12, 2–15). In North America the Great Lakes lie in glacially sculptured basins. Other types of glacial-lake basins were formed in areas with glacial deposition, such as end-

Fig. 3–13. Glacially eroded valleys in the Jotunheimen Mountains, Norway. (Photo by Fjellanger Wideröe A/S.) Observe the typical open U-shaped cross profile of the valleys, and note that the mouth of tributary valley (A) is "hanging" on the side of the main valley (B). The river from the tributary valley has eroded a gorge in the shoulder at the mouth of the valley. Note also the two lakes which lie in glacially sculptured basins on the floor of valley C.

B: Visdal Valley.

moraine zones (see p. 51). Areas that have been glaciated during the last glaciation can usually be recognized on the basis of existing lake landscapes. Outside of such areas the natural lakes are generally rare, particularly in Europe and North America (see p. 50). Many of the glacially eroded lakes are very deep; for instance, Lago Argentino in southern Argentina has a maximum depth of more than 1000 m (Fig. 3–11).

Fig. 3–14. Examples of river-eroded valleys (canyons) and ravines with V-shaped cross profiles.

A: A narrow valley or canyon cut by a river on the floor of a wide, open, glacially sculptured, U-shaped valley. B: Ravines cut by small brooks.

Fig. 3–15. Transformation of a mountain with valleys eroded by rivers and weathering processes (A) to a typical glacially sculptured mountain and valley system (C). This kind of transformation took place on most mountains which were glaciated during the late Cenozoic.

A: Valleys eroded by rivers have V-shaped cross profiles, and the floors of tributary valleys grade into the floor of the main valley at the same level.

B: One stage in the transformation of the landscape by cirque and valley glaciers.

C: Glacially sculptured valleys have U-shaped cross profiles, and trough-shaped depressions generally filled with lakes occur on their floors. Tributary valleys are hanging, and the highest mountains have an Alpine topography with cirques, peaks and sharp edges.

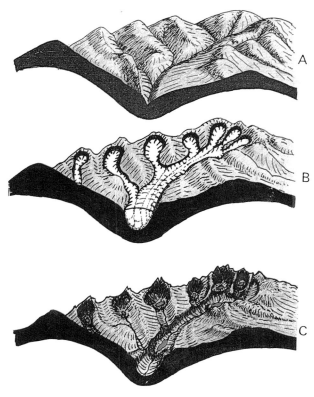

Glacially sculptured valleys

Typical glacially sculptured valleys have U-shaped cross profiles. In addition they contain rock basins on the valley floors, hanging tributary valleys, and steep cirque head-walls at the heads of the valleys (see Fig. 3–17). Therefore, glacially sculptured valleys can usually be distinguished easily from river-eroded valleys, which generally have V-shaped or narrow, box-shaped (canyon) cross profiles, and the floors of tributary valleys are more or less graded to the floor of the main valley (see Fig. 3–15).

Alpine landscapes (Figs. 1–1, 3–15 to 3–17)

Some of the most spectacular glacially sculptured parts of the world are the Alpine areas, such as the European Alps, the American Rockies and Andes, the Himalayas, and the New Zealand Alps. The highest parts of these mountains remain glaciated today, and we can observe how jagged ridges and sharp peaks are formed mainly by head-wall erosion of very active cirque and valley glaciers. Figure 3–15 illustrates how the cirque/valley glacier "eats" backwards into the mountain, supported by frost shattering at the head-wall. During ice ages the cirque and valley glaciers were larger, and many formed at much lower levels than today. Therefore, some Alpine landscapes lie beyond and below the present-

Fig. 3–16. Cirques on the crest of the Uinta Range in Utah (photo by John Shelton). The cirques were carved during one or more glaciations by small cirque glaciers which lay on the near side of an originally rather smooth mountain ridge. A considerable part of the originally smooth surface still exists on the far side and between the cirques. The presented topography represents an early phase in the formation of an Alpine topography with sharp peaks and ridges.

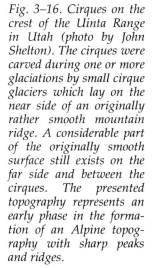

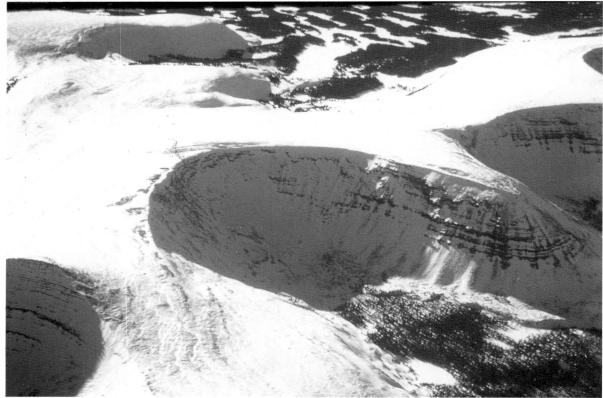

day glaciated areas. They are "fossil" Alpine landscapes.

Fig. 3–17. Cirques in a mountain massif in Jotunheimen, Norway (photo by Fjellanger Wideröe A/S). Cirque glaciers are still active in several of the cirques. Note that only small remains exist of the original smooth mountain surface. This is a more advanced stage than that shown in Fig. 3–16. A more advanced stage with typical Alpine topography is shown in the background and in Fig. 1–1.

Small glacial-erosion features

Several small-scale features also are characteristic of glacially eroded and sculptured rock surfaces. They include glacial striations, crescentic gouges and fractures, smooth stoss-slopes, and abrupt, ragged lee-slopes of projecting rock knobs, of which the whale-back knobs *(roche moutonnée)* are the most typical. These kinds of features are found within most of the glaciated regions, and they are used to reconstruct the former ice-flow directions (see Figs. 3–18, 3–19, 1–11). Very special subglacial abrasion features are the so-called P-forms (plastic forms), which are smooth, irregular, often sinuous, shallow grooves, channels and depressions (see Fig. 3–20). They are probably formed when debris-loaded water is forced under high pressure and speed through narrow subglacial channels.

Fig. 3–18. Glacially shaped whale-back rock with a gentle, glacially striated stoss-slope, and a steep lee-slope formed by glacier plucking.

Fig. 3–19. Lunate fractures (large) and crescentic fractures (in the foreground). They were formed by stress which developed in the rock when erratics in the glacier sole were dragged and forced against the rock surface. The arrow shows the glacier-flow direction.

A₂

A₁

A₃

(*Fig. 3–20, cont.*)
A₁: Pothole in Finland. (Photo by Jari Väätäinen.)
A₂: Potholes on the coast of southern Norway.
A₃: The famous pothole in Glacier Garten, Lucerne. (Photo from Glacier Garten collection.)
All of the potholes presented were eroded in hard crystalline rocks.

B: Plastic forms, P-forms, eroded by water which flowed on the glacier bed under high pressure, probably by sand-loaded water. The P-forms are best developed in front of the person. Other parts of the bedrock surface are glacially striated.

B

Fig. 3–20. Small features eroded by glacial rivers.

A: Potholes eroded by stones in the water which swirled in the pots at the foot of water falls within the glaciers. The potholes may occur on hilltops and other rock surfaces far from existing rivers.
In Scandinavia the potholes are called jättegryter ("giants pots"), since only giants could have made and used them, according to the legend.

Glacial tectonic features

The late Cenozoic glaciers, some several kilometers thick, imposed a tremendous shear and compressive stress on the ground over which they moved. As a result the stratified sediments below the glacier beds were frequently folded and thrusted, and even large slabs of bedrock and sediments could be dislocated and moved. Slabs of this kind are, for example, exposed in steep sea cliffs in Denmark. However, in some cases the glaciers moved across stratified sediments without tectonizing them. This was particularly true if the ground and the stratified sediments were frozen, as in permafrost regions, or if the glacier was cold-based and frozen to its bed. For wet-based glaciers a high hydrostatic pressure in the water film at the base, and in the pore-water of the sediments below the base, could cause the glacier to ride across the bed without disturbing the sediments significantly. Even the normal compaction and consolidation of the sediments could have been negligible if the drainage of the water in the sediments below the glacier was restricted, and consequently the hydrostatic pressure in the pore-water was high.

Field observations show that the beds which have been overridden by a glacier are frequently tectonized (see Fig. 2–23). In some cases even the glaciofluvial beds on outwash plains in front of the glacier have been folded as a result of stress from the glacier. Folds formed by an overriding glacier are generally asymmetrical and overturned, and the fold axes are oriented transversely to the ice-flow direction. Even the faults and shear planes formed by the glacier frequently have a transverse strike direction. Generally the former ice-flow direction can be determined by careful analysis of the folds and shear planes, and in Denmark, for example, these kinds of analysis have been used to develop a *kineto-stratigraphy* (see p. 190).

Some of the glaciotectonic structures are presented on p. 192, and most of them can be observed in glacially tectonized sediments.

Glacial transport

As mentioned on p. 106 and shown in Fig. 3–3, the rock and ice particles in a glacier are transported along imaginary flow lines. It is important to observe that flow lines which start at the head-wall and the side-walls in the accumulation area continue in a layer near the base of the glacier. As a consequence the rock fragments from the head-wall and side-walls will generally be transported in the same basal ice layer, together with rock fragments abraded or plucked from the glacier bed. Therefore, this layer can be loaded with rock debris. It is usually no more than 1–2 m thick, and it has been called the glacier sole (Fig. 3–24). Much of the sole material is sheared up to the glacier surface near the front of glaciers which terminate on land (see Fig. 3–21). Gradually this surface material will be transported down the steep front-slope and form an end moraine if the glacier remains internally active and the front is stationary for a long-enough period, or if the front advances.

If the glacier bed is very irregular, with projecting high and steep hills and ridges, then rock debris derived from the projecting bedrock can be transported along flow lines higher up in the glacier. In particular, if the highs project above the glacier surface as nunataks, the debris will be transported along very high-lying flow lines, and even at the glacier surface if the highs or nunataks lie near or below the equilibrium line. Tributary glaciers which join a trunk glacier at or below the equilibrium line may also carry much rock debris, which may melt out on the glacier surface in the higher part of the ablation zone. This is particularly true for material transported along the side of a tributary glacier that joins a trunk glacier to form a medial moraine with much rock debris. In this way some glaciers may transport much rock debris distributed over much of the ablation surface, and active glaciers will transport the debris to the glacier front where it is deposited, as if on a conveyor belt. However, if the climate warms and the near-front parts of the glacier stagnate, then all of the englacial rock debris will eventually melt out and accumulate on the glacier surface. Gradually this debris is lowered to the glacier bed, where it becomes an ablation moraine (see Fig. 3–28). This was a rather common situation during the late melting phase of the ice age glaciers, and it was particularly apparent during the late melting phases of some W/W glaciers.

Fig. 3–21. Transport and deposition of rock debris at ice fronts. A, B and D are fronts of temperate glaciers. C and E are fronts of polar glaciers.

A: Rock debris on the front-slope (lateral part) of Findeln Glacier in Switzerland. The debris is transported up to the glacier surface along shear planes, and it gradually slides down the front-slope and forms an end moraine.

B: Approximately the same part of Findeln Glacier a few years later, when the ice margin had retreated several meters; note, however, that the small, marked end- (lateral-) moraine ridge (x-x) was deposited before the retreat started.

C: The front of a small branch from the Beardmore Glacier, Antarctica. The debris on the glacier surface was transported to the surface along the many shear planes which are visible on the front-slope. The debris will gradually move down the slope and form an end moraine, if the glacier margin is stationary over a longer period. Note the small end-moraine ridge in front of the glacier.

D: Black shear planes where basal till is transported from the glacier sole to the glacier surface.

E: The front of a glacier in the Dry Valley area, Antarctica. The glacier is thin and frozen to its bed. Therefore, it transports no rock debris.

Fig. 3–22. The front of Myrdalsjökull Glacier on Iceland (photo by Johannes Krüger). A: The front of the glacier is greyish-black and covered with rock debris. B: The outcrop of a marked shear plane with much rock debris, which has been sheared up from the base of the glacier. M-M: A prominent young (Little Ice Age) end moraine.

Deposition of glacially transported debris, tills and moraines

Tills (Fig. 3–24B). Unsorted rock debris deposited directly by a glacier is called till, and there are several different kinds of till depending on how and where the debris is deposited.

Tills are generally composed of many different sizes of rock clasts, ranging from clay to boulders. The coarse clasts are randomly distributed within the finer-grained matrix of the till, so that the till is visually unsorted (i.e., a diamicton). Most of the larger clasts within a till are subrounded to subangular (see Figs. 3–25 and 3–59). However, a certain fraction of crushed, angular clasts often occurs together with some rounded clasts which the glacier has incorporated from subglacial sediments, such as glacial-river deposits or preglacial deposits. In some areas where a glacier overrides glaciofluvial sand and gravel beds, the till may even consist of predominantly rounded clasts.

The different kinds of till, such as lodgement till, melt-out till, flow till, deformation till, subglacial till, supraglacial till, waterlaid till, and englacial till, are all described on pp. 192–93 in the Glossary.

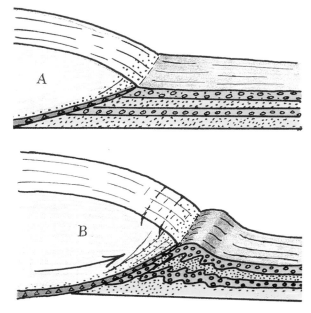

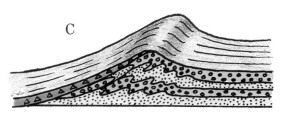

Fig. 3–23. Formation of a push moraine consisting of mainly glaciofluvial sediments. A: An outwash plain is deposited by the rivers in front of the glacier. B: The glacier advances, and the glacier front pushes into the outwash unit. C: The glacier has retreated and left an end-moraine ridge which is a push moraine.
Note that most of the moraine consists of tectonized (pushed-up) glaciofluvial sediments. A thin till bed is here left on top, but the till can in many cases be pushed and folded into the outwash unit.

Fig. 3–24A. Rock debris in transport in the sole (x) of Taylor Glacier, Antarctica. The sole is about 1 m thick. (Photo by H. Borns.)

Fig. 3–24B. Coarse-grained tills. They are all visually unsorted (diamictons) with matrix-supported large clasts. A: Basal till overlying stratified glaciofluvial (?) sand and gravel. B: Basal till. C: Basal till with clastic (till) dikes injected in the sand unit. D: Till in an end moraine consisting of much supraglacial (ablation) till.

A

B

C

D

Fig. 3–25. A typical subglacially transported, glacially striated and shaped erratic.

Fig. 3–26. Drumlin ridges on Finnmarksvidda Mountain Plateau in northern Norway. They were formed at the base of a glacier which moved towards the large lake. (Photo by Fjellanger Wideröe A/S.)
Drumlins are ridges of basal till oriented with their long axis in the ice-flow direction. They are frequently "cigar" shaped with a "tail" at the lee end, and they frequently have a core of bedrock or old deposits. Note the prominent esker ridge to the right of the largest drumlins.

Moraines. The term "moraine" is a geomorphologic-genetic term used for a topographically expressed glacial deposit. Most terrestrially deposited moraines consist mainly of till, but in some cases there can be a considerable amount of glaciofluvial sediments within terrestrial moraines. There are several kinds of moraines, depending upon how and where they were deposited.

1. *Marginal moraines* are deposited along the glacier margin. They form at altitudes lower than the equilibrium line, and they are frequently ridge-shaped (see Fig. 3–28). There are two kinds of marginal moraines, lateral moraines and end (terminal, frontal) moraines. They consist generally of melt-out and flow till transported down the front/side slope of the glacier, and of basal till or pushed-up till formed during glacier advances/fluctuations. Some marginal moraines may contain much glaciofluvial deposit. Push moraines are usually sharp end-moraine ridges composed of material pushed up at the front of the glacier during an advance. This material may consist of predominantly glaciofluvial sediments (Fig. 3–23). End moraines deposited in water will be described below.

2. *Subglacial* (basal or ground) *moraines* are formed at the base of the glacier, and they consist of blankets of lodgement and/or melt-out (ablation) till. They are deposited beneath the ice, usually in peripheral zones fairly near the glacier margin. There the friction against the glacier bed frequently exceeds the transportability of the glacier, and till is deposited by lodgement on the bed. In addition sediments in the glacier sole can melt out and be deposited as a part of the subglacial moraine, often as a blanket on the lodgement till. The subglacial moraines generally form wide, relatively thin sheets conformable with the subglacial terrain. They were formed successively as the ice front retreated and the peripheral zone migrated in proximal direction during the final W/W deglaciation. Thick subglacial moraines were commonly formed on stoss-slopes, and they are particularly well developed at the foot of some steep stoss-slopes. The stoss-slope moraines are usually very compacted while lee-slope moraines are less compacted and frequently contain some glaciofluvial material.

Drumlins (Figs. 3–26, 3–27) are streamlined ridges whose long axes lie parallel with the

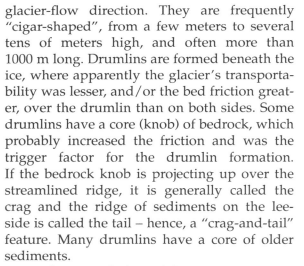

Fig. 3–27. Drumlin (A) in the drumlin field (B) in central New York. (Photo by Ward's Scientific Establishment. The map is a part of a USGS topographic map.) The drumlin was deposited at the base of a glacier which flowed in a southwesterly direction (from the upper right to the lower left corner). It has a classic shape with a long tail. Some drumlin fields in North America are very large, particularly in Canada.

glacier-flow direction. They are frequently "cigar-shaped", from a few meters to several tens of meters high, and often more than 1000 m long. Drumlins are formed beneath the ice, where apparently the glacier's transportability was lesser, and/or the bed friction greater, over the drumlin than on both sides. Some drumlins have a core (knob) of bedrock, which probably increased the friction and was the trigger factor for the drumlin formation. If the bedrock knob is projecting up over the streamlined ridge, it is generally called the crag and the ridge of sediments on the lee-side is called the tail – hence, a "crag-and-tail" feature. Many drumlins have a core of older sediments.

Another morphological feature on some areas of subglacial basal (ground) moraines are low moraine ridges, generally less than 2 m high, which lie parallel with the glacier-flow direction. These so-called "fluted moraines" are formed at the base of the glaciers near the ice front, usually by water-soaked subglacial till which was squeezed up into cavities on the glacier sole. The cavities were formed when

the glacier overrode large erratics which projected up from the glacier bed, erratics which were anchored in the frozen bed below the thawed, water-soaked till. In some cases the fluted surfaces, with both low and higher parallel ridges, were formed by glacial erosion, and even glacially sculptured bedrock surfaces can be fluted.

3. *The supraglacial (ablation) moraines* consist of melt-out (ablation) till and flow till formed when rock debris on the surface of a stagnant part of a glacier is lowered down to the glacier bed by melting of the ice. The way rock debris is brought up to the glacier surface in the ablation zone of an active glacier and the way an active glacier becomes stagnant are described on p. 119. Stagnant, near-front parts (distal parts) of glaciers can be observed on several recent, well-known glaciers. The Tasman Glacier in New Zealand is an excellent example (Fig. 3–28); there, an irregularly thick cover of supraglacial (ablation) till rests on the melting, stagnant remnants of the distal parts of the glacier. Differential melting results in a very irregular, bumpy surface topography, where till on the high parts flows down into the depressions. The result, when all ice has melted, is a hummocky topography which is typical for many supraglacial (ablation) moraines (see Fig. 2–62). Very hummocky ablation moraines are frequently called dead-ice moraines.

A

B

C

Fig. 3–28. The stagnated frontal part of Tasman Glacier, New Zealand. A: Looking upglacier, with the active white glacier in the upper left background. B: Looking downglacier. C: Aerial view, looking downglacier (photo by H. Borns). x: Photographs A and B were taken from this point.

A veneer of ablation (supraglacial) till covers the surface of the stagnated section of the glacier. The thickness of glacier ice below this veneer varies from a few to several tens of meters. The hummocky topography is caused by differential melting of the ice. Ablation till slides down into the depressions, where it can be several meters thick. The final result, when all ice has melted, is a hummocky topography, which is typical for many ablation moraines, and is frequently called a dead-ice topography (see Fig. 2–62).

Picture A was taken in the spring (late October) and B three months later. The pictures show that small changes took place in three months, such as the formation of a small pond in front of the black dot in the lower left corner of picture B.

Lateral moraines border the glacier on both sides of the valley. They rise up to 30–50 m above the glacier surface. As late as 1930, an active Tasman Glacier filled the valley up to the crest of the lateral moraines.

Fig. 3–29A: Cross section of the central and distal parts of a marine-deposited end-moraine ridge in southeastern Norway.

Distal section: Foreset beds (A) of sand and gravel which dip about 15° in distal direction.

Central section: Till beds (I, II_1 and II_2) which include strongly folded beds and lenses of sand and gravel. Foreset beds (A_1, A_2 and A_3) which are tectonized and folded.

Proximal section: Till beds (I, II and III) interbedded with stratified glaciofluvial, gravel, cobble and boulder beds (B^1 and B^2) which dip in proximal direction. They were supposedly deposited by subglacial rivers which flowed in ice tunnels, under hydrostatic pressure, up the proximal moraine slope (see Fig. 3–29B). However, the beds could have been deposited on top of ice which melted and thus tilted them.

All beds were deposited 50–100 m below sea level, and marine clay and beach gravel lie on top of unit I in another part of the pit.

Fig. 3–29B: Cross section through an end moraine which is being deposited below sea level at a grounded ice front.

Dots: sand and gravel. Triangles: erratics in till and in the basal part of the glacier. Black: glaciomarine silt and clay.

A: Bottomset unit. B: Foreset "fan" unit, consisting of sand and gravel beds deposited mainly by gravitational sliding, as in fans and deltas. C: Till, including flow till, basal till and ablation till. D: Glaciofluvial unit, supposedly deposited by subglacial rivers which flowed uphill in ice tunnels under hydrostatic pressure; conversely, the beds may have been deposited on top of ice which later melted and tilted the beds.

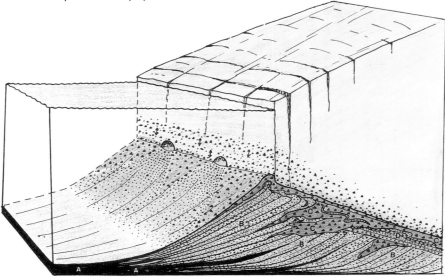

Marine-deposited end moraines and other glacio-marine sediments

Marine end moraines

Numerous end moraines were deposited in the sea along grounded ice fronts during the melting phase of the ice sheets in the coastal districts of North America and northern Europe. They are generally stratified, and consist of till units and a variety of glaciomarine stratified deposits, including deltaic-type foreset or fan beds. Today many moraines of this kind can be observed on land along isostatically emerged coasts, and they often have a thin cover of marine clay or shore deposits, formed during their emergence (see the discussion on p. 16). Many of the marine end moraines are very distinctive ridges (see Figs. 3–29A, 3–29B). Studies of the deposits within some ridges indicate that the moraines were built at active oscillating ice fronts which were alternately afloat and grounded.

Grounding-line moraines (Fig. 3–6)

A grounding-line moraine is deposited at the line or zone where a glacier, generally a polar glacier, starts to float and grade into an ice shelf. Along a grounding-line, or zone, the glacier may have contact with the bed in some periods, and be barely floating at other times. Since a glacier usually carries most of its rock debris in the basal part, in the sole, much of this debris melts out when the sole comes in

contact with the water. In this way considerable amounts of debris are deposited in the grounding-line zone. However, part of the debris-laden sole usually continues to move into the floating ice shelf, where it finally melts through contact with the water, well seaward of the grounding-line, or it may continue in icebergs which break off from the front. However, icebergs which break off from the large polar ice shelves are usually fairly clean, with little rock debris at the base. Since most polar glaciers produce little meltwater, the transport of material with subglacial rivers is generally small. Even the volume of glacial debris transported within the glacier, or in the glacier sole, is usually small, and therefore the accumulation at the grounding-line can be very slow. However, not all material which is deposited at the grounding-line moraine necessarily melts out from the glacier sole. Some material from beneath the glacier is clearly dragged and pushed into the moraine zone.

Considering the gradual melt-out and deposition of rock debris in relatively wide grounding-line zones, the grounding-line moraines of polar glaciers may be wide, low ridges, or there may be no sharply defined ridges at all, just small breaks in slope. The grounding-line moraine consists mainly of beds that are dipping gently in a distal direction, and the sediments within the beds are often diamictons (glaciomarine diamictons or flow tills), especially on the distal side. The proximal part of the moraine may be more complexly deformed and contain a higher amount of basal till. The volume of glacial debris deposited at the grounding-lines of subpolar and temperate glaciers can be considerable, and more distinc-

tive moraine ridges can be formed. They are usually a combination of grounding-line moraines and end moraines, since the ice shelves at these types of glaciers are very narrow and, in part, missing. In fact many of the moraines described as marine-deposited end moraines are probably of this kind (see p. 126).

Other glaciomarine deposits

Figure 3–6 presents the glaciomarine zonation used in this book. The clay usually dominates in all zones, and the amount of ice-dropped coarser material, including large dropstones, generally decreases away from the grounding-line moraine via the ice-shelf zones to the distal iceberg zone, or from the ice-proximal to the ice-distal zone. Marine life also changes in the same direction. Only very specialized high-polar organisms can live in the cold, dark water under the thick ice shelf, and other cold-water communities live in the ice-proximal zone or proximal iceberg zone, which is normally covered with sea ice during the winter (see p. 108).

Several different definitions have been presented in previous literature for the term "glaciomarine deposits". However, all scientists agree that at least the deposits in the ice-shelf zone, the proximal iceberg zone and the ice-proximal zone are glaciomarine, and many consider the deposits in the distal iceberg zone and the ice-distal zone to be glaciomarine.

The glaciomarine deposits are predominantly composed of clays and silts with a variable content of ice-dropped coarse clasts, and they generally contain a cold-water fauna. Much of the clay and silt particles were transported in suspension in the water column, while some were transported and deposited by turbidity currents and other bottom currents. Clasts which are liberated from the ice front by melting and come to rest at the toe of the grounded ice front may subsequently move distally with submarine mud flows or turbidities.

The annual accumulation rate near the grounding-line of polar glaciers is generally very low, from a few centimeters to a few millimeters. However, accumulation rates near the grounded ice fronts of temperate and subpolar glaciers can be much higher, and annual rates of 4–9 m have been recorded in Alaska.

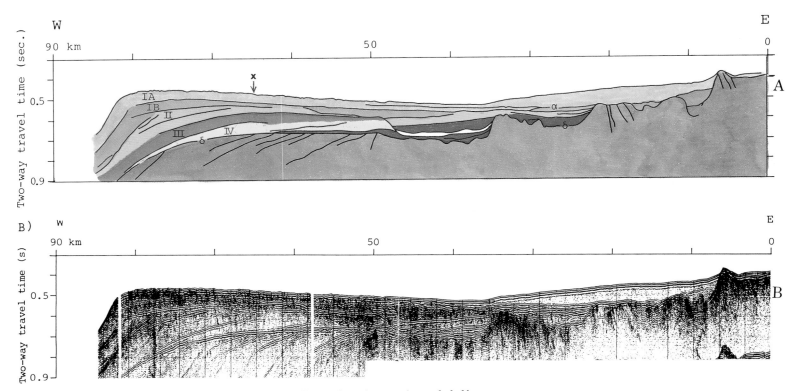

Fig. 3–30. A seismic sparker profile (A) across the shallow submarine continental shelf off Spitsbergen, and the interpreted seismic units (B). (Modified from Kristoffersen et al., 1981.) All units, I to IV, represent late Cenozoic sedimentary beds. The total thickness of beds I to IV below x is about 200 m. The irregular bedrock surface (δ) is a marked erosional surface on top of pre-late-Cenozoic sedimentary bedrock (brown). Similar profiles exist for many shelf areas along the previously glaciated coasts both in Europe and North America. The "texture" of the seismic signal from the units frequently gives a good indication of what kind of sediments are present, but usually sedimentological studies of cores from the units are needed for more precise interpretations.

There, stratified units of mainly clay, silt, and sand are being deposited. Even coarse gravels are frequently deposited at the grounded fronts of temperate and subpolar glaciers (see Fig. 3–29A).

Laminated clay/silt units, which generally include some sand lamina, are usually deposited fairly near the ice front. In many cases the sand is transported by jet-currents extending from the submarine mouths of rivers in ice tunnels at points along the margin. More distally from the ice front, and even in some areas closer to the front, the glaciomarine clays are frequently not clearly stratified. Most glaciomarine clays which contain coarse-size grains are poorly sorted, and they can be classified as diamictons. They may consist of material ranging from sand to boulders, generally ice-dropped, together with the finer-grained clay and silt particles. Observations in the Arctic show that some of the glaciomarine sediments even contain portions of fine-grained silt and clay, which have been transported with the sea ice. The volume of coarse-grained sediment deposited in the distal-iceberg/ice-distal zones varies, but it is usually very low.

Seismic profiles from the sea floor on the continental shelves along the coasts of North America and northwestern Europe display many seismic units or beds, while sediment cores from the shelves show that many of the units are diamictons composed of glaciomarine clays, clayey tills, and till-like sediments. Frequently the identification of the true tills has been problematic, but in many cases an overconsolidation of the sediments, a mixed-fauna population, and a preferred orientation of the coarse clasts can be used as criteria.

Iceberg ploughing

The icebergs which drifted in the iceberg zones frequently reached and ploughed the shallow parts of the sea floor. Consequently they tectonized (mixed and disturbed) the glaciomarine deposits. This was very much the case with glaciomarine deposits on shallow continental shelves along the coasts of North America and northern Europe. The numerous fossil plough marks observed on sections of the shelves are often parallel, as in ploughed fields, and they clearly demonstrate the importance of iceberg ploughing.

Quick clays and clay slides

Suspended clay particles in marine water are slightly electrically charged, and they can be pulled together in the saline sea-water, where they flocculate and form aggregated particles which are heavier than single clay particles. Aggregates thus formed, and formed in other ways, drop relatively rapidly to the sea floor together with single-grain clasts of about equal weight. This process produces marine deposits showing no clear, visible bedding. The flocculated clay deposits have a very open structure and hence a large pore volume. When this clay is raised above sea level, as it was in many isostatically uplifted, formerly glaciated areas, the stabilizing salt ions may gradually be leached out of the clay by groundwater, causing the electrical bindings between the clay particles to weaken or disappear. This kind of sediment, saturated with water, is called "quick clay". Subjected to deformation by stress, the open clay structure may collapse, making the clay a "thin soup" which flows almost like water, due to the high content of pore-water. Numerous clay slides have occurred in this way, and many of them have caused considerable damage, for example, the famous slides near Trondheim in western Norway: the Verdal slide in 1893, in which numerous farms were destroyed and 112 people killed; and the Rissa slide in 1978, which took 5–6 million m³ of ground and destroyed 7 farms, but during which only one person was killed. Clay slides are common also in many other areas with emerged glaciomarine clay deposits, such as the Oslo fjord region in southern Norway, the west coast of Sweden, the St. Lawrence Lowland of Quebec in Canada, and the coast of Maine in USA.

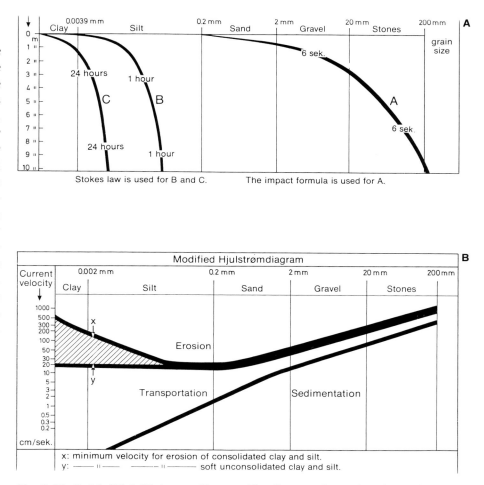

Fig. 3–31. B: Modified Hjulström diagram. The diagram shows the relation between water-flow velocity and erosion, transport and deposition of grains of different sizes, and erosion of different sediments. The limit for the erosion zone varies between line x and line y, depending upon the character (consolidation) of the clay and silt sediment. Note that consolidated clay is hard to erode, and that coarse-grained silt to fine-grained sand is most easily eroded. Note also that clay and fine-grained silts are being transported, and not deposited, even with flow velocities less than 0.2 cm/s.

A: Illustration of approximately how fast different-sized particles drop in water. A: distance dropped in 6 s. B: distance dropped in 1 h. C: distance dropped in 24 hs. Note that a 200 mm particle drops about 10 m in 6 s, a medium-sized (0.063 mm) silt particle drops about 10 m in 1 h, and clay particles drop less than 1.5 m in 24 hs.

Glacial rivers and lakes; erosion, transport and deposition

General remarks

A river transports a bed load, a suspended load, and a dissolved load. The bed load consists of large, rolling grains and the somewhat smaller, saltating (bouncing) grains. The size of the grains in the two categories varies with the current velocity. Silt and clay grains usually constitute the suspended load, which is generally deposited where the current velocity approaches zero, such as in lake and in ocean basins. However, grain-size analyses of river sand, for instance, may reveal small portions of silt/clay and gravel mixed into the sand. They supposedly represent suspension and rolling members, respectively.

The modified *Hjulström diagram* in Fig. 3–31 shows the current velocities needed to erode, transport and deposit clasts of various grain sizes. Note that the minimum velocity needed

Fig. 3–32. Water gates in the fronts of Fox Glacier and Franz Josef Glacier in New Zealand. The gates are the openings of subglacial ice tunnels which are kept open by subglacial rivers. Note the bouldery gravel ridge which is being deposited by the river in the picture to the right. The ridge probably continues into the tunnel, and in that case it is a part of an esker ridge which is extremely bouldery.

Fig. 3–33. The opening of an abandoned ice tunnel formed by a glacial river which flowed at the base of the Matanuska Glacier, Alaska. Note the river-transported gravel on the floor of the tunnel. This gravel can be left as an esker ridge on the landscape following the deglaciation. (Photo by Edward Evenson.)

to erode a sediment of one grain size is considerably higher than the minimum velocity needed to keep the grains of the same size in transport. This is particularly true for grain sizes smaller than coarse sand, and especially true for clay particles. Therefore, when the particles come to rest on the river or sea floor, a considerably higher velocity (energy) is generally needed to put them in motion again. For instance, a 0.03 mm silt particle can settle at a current velocity of about 0.2 cm/s, but a velocity of 20 cm/s is needed to move a particle of this size from a normal sediment on the river bed. The velocity at which clay and silt particles finer than 0.01 mm can settle is set at 0 in the diagram, but velocities between 50 cm/s and 200 cm/s are needed to erode normal sediments with this kind of particle. Note also that median sand is the most easily eroded sediment.

Settling velocities of sediments

The settling velocities in water for the different-sized particles shown in the Hjulström diagram can be more easily understood by studying Fig. 3–31A, where the dropping times and distances for particles of various sizes are shown. Note that fine-grained silt and clay particles drop a very short distance in 1 hour. Clay-sized particles, for example, drop less than about 1 m in 24 hours. Therefore, the slightest turbulence and current flow of water will, in general, prevent grains of that size from settling.

Glacial river (glaciofluvial) deposits

Glacial rivers characteristically experience great variation in discharge and in volume and grain sizes of the sediments that they transport. Rivers draining glaciers are frequently fully loaded with sediments, which means that they carry as much sediment as possible at a given velocity. As an overall result they are aggrading rivers, and therefore have a braided pattern of channels. In cross section the deposits display a cut-and-fill bedding.

The base of most glaciers is a "grinding mill" which produces a wide range of particle sizes, from boulders to clay. Much of this sediment is fed from the ice into meltwater rivers together with debris carried in englacial and supraglacial positions. Consequently glacial rivers transport much clay, silt, and fine sand in suspension, which colors the water "milky white", while coarser clasts are generally transported on or near the bed by rolling, sliding or bouncing (saltation). During periods with much melting and rain, the water discharge from the ice can be very high, and large boulders may be transported. During the melting phase of both the North European and the North American ice sheets, several ice-dammed lakes drained in a catastrophic fashion, and this led to floods which scarred the landscape (Figs. 2–45, 2–65) and transported boulders as large as small houses. In other periods, particularly during winter, when the melting and water discharge were low, the suspended load dominated, and mainly finer-grained sediments were deposited.

Due to the large seasonal variations in discharge and current velocity of glacial rivers, the variation in size of clasts which are transported and deposited on the beds is usually considerable. As a consequence units of glaciofluvial deposits are stratified, and the grain size may vary widely from one bed to the other. This is particularly true for units deposited close to the glacier. Units that are deposited in contact with the glacier, the ice-contact deposits, can be poorly stratified and poorly sorted. In fact there are ice-contact glaciofluvial deposits in which clasts that originated from englacial or supraglacial debris have been transported only a few meters by the water. This kind of deposit may resemble till, except that it is faintly stratified and has lost most of the fine grains. Also, the stones may have the same degree of "rounding" as stones in tills. However, stones that have been transported longer distances with the glacial rivers are better rounded and some even well rounded (see p. 153). Deposits transported a long distance from the glacier are generally well sorted also. All river deposits, including glacial river deposits, are clast-supported rather than matrix-supported (see p. 155).

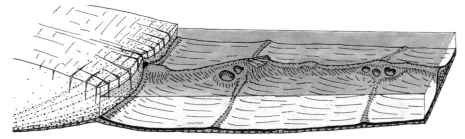

Fig. 3–34. The formation of a "subaquatic" esker, a De Geer Esker, also called a hummock esker ("punktaasar" in Swedish). The esker sediments are deposited in a body of standing water, generally a lake, at the mouth of a subglacial river, and segments of the esker ridge are formed successively as the ice front retreats. The hummocks represent phases when the ice front was stationary for a longer period. They may correspond with small end moraines as indicated, but moraines are frequently missing. In some cases the hummocks have flat-topped surfaces which may correspond with the lake or sea level, and then they are true small deltas. The accumulation of esker sediments may take place both at the mouth of the tunnel and within the adjacent part of the tunnel. In many cases the hummocks are poorly developed or missing. Figure 3–35 illustrates the formation of a subglacial and a subaerial esker.

Figures 3–32 to 3–39 present some of the most important glacial river features.

Eskers

These are ridges of stratified sediments deposited by glacial rivers in tunnels underneath the glacier, in open channels within the glacier, or at the mouths of glacial rivers which terminate at ice fronts resting in lake or ocean water. They are called subglacial, subaerial and subaquatic eskers, respectively. Patches of till and scattered large erratics are commonly found on top of, or even within, the eskers, while the subglacial eskers may have a more or less continuous cover of ablation till. They are frequently long, sinuous ridges. Subaquatic eskers are probably deposited both within the glacier tunnel adjacent to the outlet and in the open water at the mouth of the tunnel, in front of the retreating glacier (see Fig. 3–34). A hummock of river sediments may be formed at the mouth, a hummock which consists mainly of submarine/sublacustrine fan beds deposited in the open water when the ice front is stationary. When built up to sea level the hummock becomes a flat-topped delta. Eskers with hummocks deposited in open water have been called "hummock eskers" ("punktåsar") or "De Geer Eskers" after the Swedish geologist who explained their origin. In some cases the hummocks are well developed, but frequently they are missing. Kettle holes and other collapse

Fig. 3–35. Formation of different glacial river (glaciofluvial) features.

A: A stagnant valley glacier (white) with glacial rivers flowing in tunnels below the glacier, in open "canyons" on the glacier, and on a terrace along the side of the glacier, between the glacier and the valley side. In addition, a delta is deposited and a shoreline is formed in an ice-dammed lake (on the left side).

B: Features exposed when the glacier has melted. E: subglacial esker. E₁: subaerial esker. K: kames. KT: kame terrace. D: glacial-lake delta. S: glacial-lake shoreline. X: basal till. Y: ablation till.

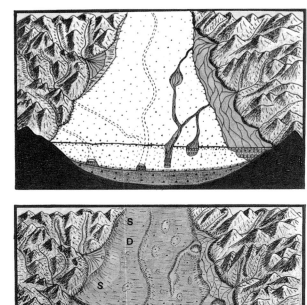

features and structures caused by the melting of ice buried beneath the sediments are commonly found along with eskers and associated hummocks.

Kames

These are knolls, frequently circular in plan, composed of glaciofluvial sediments which are deposited in basins on top of, within or below the ice (see Fig. 3–35). They generally display many of the same types of sedimentary and collapse features as are associated with eskers.

Fig. 3–36. Deposition of an ice-contact outwash plain and terrace, and the formation of a dead-ice terrain at a stagnant glacier.

OP: outwash plain. O: outwash gravel and sand beds. D: dead-ice terrain with eskers, kames and kettle holes (K). C: glacial meltwater river gorge. S: subglacial river channels (dashed lines). I: dead ice covered with glaciofluvial sediments. Kettle holes (K) are formed when the ice melts. A: melt-out (ablation) till. G: glaciofluvial sediments. B: older basal till. Note the glaciotectonic collapse structures on the ice-contact slope.

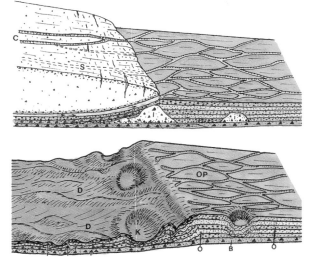

Kame terraces

These lie along valley sides, and they consist of sediments deposited by glacial rivers which flowed along the sides of dynamically dead valley glaciers. Collapse structures and kettle holes are frequently developed on the ice-contact slopes of the kame terraces (see Fig. 3–35).

Outwash plains (Fig. 3–36)

These are formed by glacial rivers flowing beyond the glacier front. The heavily sediment-loaded glacial rivers are unable to transport all material when they traverse the plain in front of the glaciers, where the flow speed is usually drastically reduced. This, in addition to the seasonal, rapidly changing water discharge, results in deposition of the bed load in the river channels, which may be filled rapidly with sediments. In this way the rivers are often forced to overflow the river banks, change course, and form braided river systems that characterize the channel pattern of outwash plains. Another typical feature is the rapidly decreasing size of the clasts in distal direction on the plain, and a corresponding rapid decrease in slope-angle of the plain. The slope-angle is naturally adjusted to allow the rivers to transport the bed load, and it must be steep near the glacier where the largest clasts, such as boulders, are transported.

Outwash plains start at an ice-contact heads, which is often an end moraine, but end moraines are sometimes missing, and the plain may start at an ice-contact zone, or an ice-contact slope where the deposits are primarily stratified and often display an ice-contact

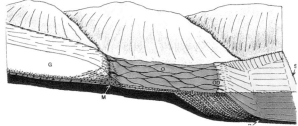

Fig. 3–37. Deposition of an outwash delta. G: glacier. M: end moraine. O: outwash plain. OD: outwash delta. S: sea level. T: topset beds. F: foreset beds. B: bottomset beds. In some cases an ice-contact slope with no end moraine is formed at the proximal end of the outwash plain (see Fig. 3–36).

Fig. 3–38. An outwash plain and an outwash delta in front of a glacier lobe from the Greenland Ice Sheet in western Greenland. (Photo by Henrik Höjmark Thomsen.)
Note the braided river pattern, which is characteristic for glacial rivers on outwash plains.

topography with kettle holes. Long, narrow outwash plains confined to valleys are called "valley trains".

characteristic of areas where stagnant glaciers melted (see Figs. 2–62, 3–28).

Outwash deltas (Figs. 3–37, 3–38)

These are deposited by glacial rivers, and their delta plains are outwash plains. Most outwash deltas have steep front-slopes, which frequently dip from 10° to 30°. Such outwash deltas are generally called *Gilbert deltas*. They have well-defined topset, foreset and bottomset beds. The topset unit usually consists of beds with very coarse-grained sediments, and it rests unconformably on the foreset unit. This unconformity was formed by erosion of the rivers during periods with high water discharge and strong river currents, frequently at or slightly below the sea or lake level. The foreset beds grade into the bottomset beds, which generally consist of fine-grained sediments, frequently silt or clay.

Dead-ice topography

This is a hummocky topography formed partly by glaciofluvial elements such as kames, small eskers, and kettle depressions, and by supra-glacial (ablation) moraine. The topography is

Ice-dammed lakes

In some areas the margins of the large ice sheets in Europe and in North America lay in terrain which sloped towards the ice, and the ice fronts blocked the normal drainage. In this way both small and large ice-dammed lakes were formed. Some lakes existed in areas where there are no lakes today, for instance, in USA, Canada, and the former Soviet Union (see Figs. 2–19, 2–57). However, in other areas, like the Great Lakes region and the Baltic Sea region, the ice-dammed lakes occupied depressions where there are lakes or ocean water today, but the glacial lakes often extended far beyond the present-day shores. Smaller ice-dammed lakes were formed in the Scandinavian high-mountain valleys during the final melting phase of the ice sheet (see Fig. 2–67).

Fine-grained sediments deposited in the ice-dammed lakes are frequently laminated. Much water poured into the lakes from glacial rivers during the summer season, carrying a wide range of grain sizes. The coarse grains settled out and formed a thick summer lamina, while the clay remained in suspension due to the

Fig. 3–39. Kettle holes with kettle lakes on the Pineo Ridge ice-contact delta in Maine, USA (see Fig. 2–48). The kettles were formed when large ice blocks melted – blocks which were buried below sediments transported by glacial meltwater rivers. X: the delta front. (Photo by H. Borns.)

melted and eroded channels through the damming ice, and suddenly the lakes emptied. In some cases the resulting floods caused much damage and eroded deep canyons, even in hard bedrock. There are numerous examples of this kind of erosion, for instance, Jutulhogget in the Scandinavian mountains, and the Scabland in Washington State, USA (see Figs. 2–65, 2–45). In Iceland the melting of glacier ice caused by heat from volcanic activity resulted in similar catastrophic floods, which are called Jökulhlaup.

summer turbulence. Subsequently, during the winter season when the lakes were frozen, much less water flowed into the lakes and the turbulence diminished. Therefore, the clay, which had been held in suspension, could settle and form a thin winter clay lamina. A pair of summer and winter beds/laminae represents an annual-couplet layer, which is usually called a varve (see Fig. 3–57). In the Baltic region a time scale back to about 13 000 years ago is recorded in the observed varves. Coarse-grained clasts like pebbles and cobbles dropped from floating ice often occur within the laminated units, but varve sequences devoid of clearly ice-dropped clasts are also common. Areas with Late Weichselian/Wisconsin ice-dammed lake deposits are now frequently excellent farmland (see Fig. 2–59).

Ice-dammed lake shorelines and deltas may be well developed, since the glacial rivers usually carried much coarse material which was deposited in the deltas, and freezing and thawing along the shores of the fresh-water glacial lakes favored shoreline development. The shorelines are usually slightly tilted due to the glacio-isostatic rebound. This is particularly true for the shorelines in the Baltic and Great Lakes regions (see pp. 17, 18).

Catastrophic drainage of ice-dammed lakes (Jökulhlaup)

Many of the ice-dammed lakes drained catastrophically. The lake basins were filled with water during the summer season; the water

Periglacial features and deposits

Wind-transported (eolian) deposits (fierce katabatic winds)

One of the most characteristic phenomena in areas adjacent to the large ice sheets is the periodically strong katabatic wind. Cold heavy air from the high-pressure areas centered over the ice sheets may pick up fierce velocity when it flows downslope, under gravity, towards the marginal zones. Velocities of close to 500 km per hour have been measured in Antarctica, where even pebbles and cobbles are transported by the strongest wind.

The cold, high-pressure air over the ice-age ice sheets was very dry, and it became even relatively drier as it flowed down the glacier slope and warmed adiabatically. As a result of these strong, drying katabatic winds, some areas adjacent to the ice-sheet margins became extremely dry, and sand, silt and clay particles could easily be blown from the dried-out surface of outwash plains, for example. Sand was usually transported along the ground surface and deposited in beds and dunes when the wind speed was reduced, generally not very far from the glacier. The cover-sand in parts of northern Europe is a good example. However, silt particles together with some clay and fine sand were transported in suspension as dust clouds higher up in the air, and they dropped out over wide areas as silt blankets, the so-called loess sheets. In addition to the katabatic winds, strong cyclonic and anticyclonic winds also helped in distributing the sand and loess (see pp. 66, 71). The loess sheets which were deposited during the ice ages cover large parts of Europe, North America and several other periglacial areas. They have since developed into some of the world's best farming soils. Extensive loess sheets in central China also were deposited mainly during the glaciations (see p. 22).

Loess units, several tens of meters thick and containing many loess beds, exist in several areas of the world. Each loess bed usually represents a cold phase (a glacial or stadial), and soil horizons between the beds represent the interglacial or interstadial subaerial weathering and soil formation which occurred on top of each loess bed.

Some of the finest-grained dust particles were transported up into the higher atmosphere and spread by jet-streams over very wide areas, far beyond the true loess districts. They are even observed as "dust layers" in glacier ice cores from Antarctica and Greenland (see Fig. 2–36).

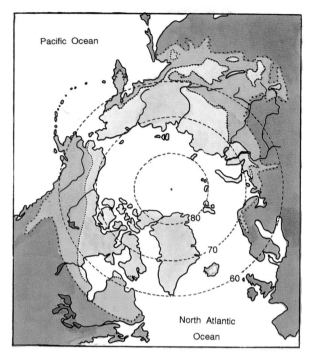

Fig. 3–40. Permafrost zones. Continuous permafrost (blue) and discontinuous permafrost (orange). (Modified from various sources including T. Pewé, personal communication.)

Frost features in the periglacial zones

The present-day periglacial (near-ice) zones lie either at very high altitudes or at high latitudes. However, during the ice ages when ice margins were located at New York in North America and to the south of Berlin in Europe, the corresponding periglacial zones lay far to the south, and reached near the Mediterranean in Europe and to Virginia in North America. The periglacial zone consists of a tundra zone with continuous permafrost and, beyond that, a zone with discontinuous permafrost. To the south of the present permafrost zones in the northern hemisphere, there is a zone with strong winter frost. The present-day zonation is outlined in Fig. 3–40, and the ice age zones are shown in Figs. 2–19, 2–31 and 2–33. Special frost features characterize the different zones. Many features are present in all of the zones, but to a different degree. However, some features are unique to the former permafrost zones, such as the ice-wedge and sand-wedge casts.

Some typical frost features are shown in Figs. 3–41 to 3–46. The principal mechanisms behind the formation of most features is the 9% volume expansion when water freezes to form ice, and the contraction of the ground ice, with resulting cracking, when the cooling continues

to lower freezing temperatures. Contraction cracks will frequently form in frozen soil when the ground temperature drops below minus 15° to minus 18°C.

The active layer

In some literature the "active" layer is defined as the upper layer in the permafrost zone which thaws in the summer. However, in this book the term will be used in a wider sense for the upper layer of the ground which freezes in the winter and thaws in the summer, even in areas outside the permafrost zone.

Expansion by freezing of the active layer

The volume of the active layer increases and the surface rises when the water within the layer freezes. Fine-grained sediments generally expand most, the clay because it has a high pore volume which is filled with water, and the silt mainly because it has a high capillarity and can hold and draw water which freezes to ice layers and lenses. The coarser-grained sediments, like sand and gravel, are commonly well drained and contain little water in the upper layers. However, in the permafrost zone where the ice below the active zone prevents

A

B   C

Fig. 3–41. Blockfields formed by periglacial frost action which breaks up the bedrock.
A: Blockfield (Felsenmeer) on Blaahö Mountain in central Norway. The top lies 1600 m above sea level, and it was a nunatak during much of the Weichselian Glaciation. The blockfield was formed mainly during nunatak phases. Lake Vaagaavann is in the background. B: A fossil blockfield (Felsenmeer) in the southern Appalachian Mountains formed during ice age maxima. C: Large quantities of frost-shattered rock fragments cover the mountain slope in Willow Creek Valley, Alaska. Some of the rocks were transported by small glaciers.

the drainage downwards, even the coarse-grained sediments may contain much water, which causes expansion on freezing.

Frost heaving of stones in the active zone is a well-known phenomenon, particularly among farmers in cold regions, who each year observe new stones appearing on the surfaces of their fields, as if they had grown there. They were lifted up with the active layer during the winter and did not drop down to their original positions when the layer thawed in the spring. Two different theories have been proposed to explain how this happens. According to one theory the stones are pushed up by the ice, and according to the other theory they are pulled up. (See frost features in the Glossary.) The stones travel toward the freeze surface, which is generally parallel with the top surface. Flat or elongated stones tend to turn to a near-vertical position (with the long axes vertical) during the process.

Ice wedges and sand wedges (Figs. 3–42 to 3–44)

As already mentioned, the low temperatures may cause the ground to contract and crack, and with repeated contractions the crack may expand to become a wedge. This repeated growth occurs only within the permafrost zone, and usually the wedge is filled with ice. It is an ice wedge, which can grow to considerable dimensions. The top of some wedges are several meters across, and they can grow to be more than 10 meters deep. However, smaller dimensions are more usual. When the ice melts the wedge will be filled with rock debris which drops from the side-walls and the top, and it becomes an ice-wedge cast. Cracks and wedges formed in some tundra and polar desert regions may be filled with aeolian sand instead of ice. They are sand wedges. Since true ice wedges and sand wedges are formed only in permafrost regions where the mean annual temperature is a minimum of minus 6° to minus 8°C, they are good climate indicators. However, in some cases it can be difficult to distinguish between well-developed frost cracks and weakly developed true ice (sand) wedge casts.

Ground ice and thermocarst (Fig. 3–43)

More or less irregular bodies or layers of "pure" ice may form within the ground in permafrost regions, and they may obtain considerable dimensions. When this ice melts, or part of it melts, the surface becomes bumpy with depressions and hummocks. It becomes a thermocarst area. Recent studies indicate that some of the large ground-ice bodies in Siberia could be remnants of old glacier ice.

Blockfields (Felsenmeer, frost-shattered rock surfaces) (Fig. 3–41)

A Felsenmeer, also called a blockfield, forms when water freezes in cracks and fissures in the bedrock and breaks it up into a jumble of angular rock blocks and smaller rock fragments. Fluctuating temperatures of considerable amplitudes, which permit the water to freeze and thaw at levels below the bedrock surface, are necessary to form a Felsenmeer.

Patterned ground

Elaborate systems are developed to classify the variety of patterned ground features, but only a brief outline of some features will be mentioned here. Most of them fall in one of the following categories: (a) polygons and related features; (b) mounds of various kinds; and (c) solifluction or congelifluction features caused by the downslope movement of the active layer.

Polygons and circles (Fig. 3–45) with and without concentrations of stones in their crack zones (sorted and unsorted) are some of the most common frost features. The reason why unsorted polygons and circles develop is either that the sediment consists of only one grain size, or that water is missing in the active layer, and therefore no freeze-thaw process and

Fig. 3–42. Ice wedge in Siberia. (Photo by Andrei V. Sher.)

The beds in contact with the wedge are bent up. However, they will be partly faulted and bent down into the cast which forms when the ice melts.

Fig. 3–43. Ground ice and ice wedge near Fairbanks, Alaska. When the protecting soil/vegetation cover is removed by ploughing, the exposed ice gradually melts and changes the field to a hummocky thermocarst area. x below two persons.

Fig. 3–44A. Sand-wedge cast polygons in Montana, USA. They were initiated by shrinkage and cracking of the frozen ground in Wisconsin time, and the cracks expanded by repeated shrinkage of the ground during the coldest Wisconsin phases. As the cracks opened they were gradually filled with wind-blown (eolian) sand (see Fig. 3–44B). (Photo by Brainead Mears.)

Fig. 3–44B. Cross sections through a sand-wedge cast within the same area as the casts in Fig. 3–44A. Note that the beds adjacent to the wedge are bent up. Apparently this is typical for most sand-wedge casts, and it is probably caused by the drop-down of the beds when the ice melts in the frost-heaved ground next to the wedge. (Photo by Brainead Mears.)

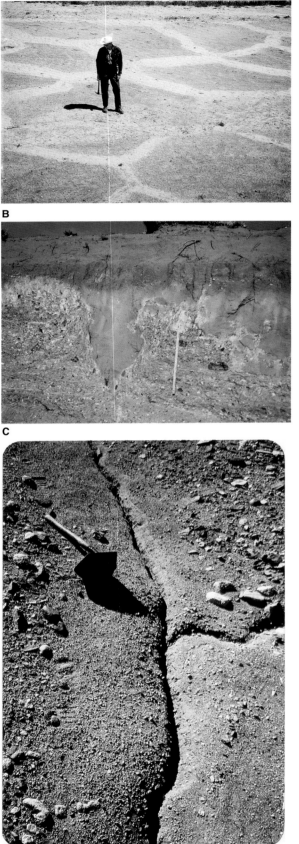

Fig. 3–45. Frost polygons in Greenland (A), Antarctica (B+C) and Iceland (D). A: sorted polygons. B, C, and D: unsorted polygons. E: stone stripes.

hence no transport of stones take place, as for instance in some polar deserts. The stones in stone polygons (sorted polygons) were generally first transported up to the top surface (frost-heaved), and afterwards laterally to the cracks. However, some stones may travel

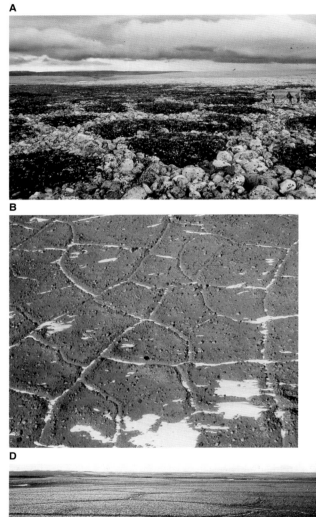

directly to the cracks, if the freeze surface is bent down over the cracks. Stone "polygons" formed on gentle slopes frequently become elongated due to a downslope movement, and the elongation increases with increasing gradient of the slopes. The final result on the steeper parts of the slopes is stone stripes or block stripes. Much fines may remain in the center of some polygons/circles as a result of the sorting process, and there ice lenses can readily form and raise the centers.

The various types of *frost mounds* are described in the Glossary. There are large and small mound types, and mounds with and without ice cores or ice lenses. Pingos are inorganic ice-cored mounds which can be 20–30 m high, and palsar are organic ice-cored mounds formed in bogs. They are generally no more than 5 m high (see Fig. 3–46). Both types form in zones with sporadic permafrost, and pingos in areas with thin, continuous permafrost also. Small mounds, less than 1 m high, can form in permafrost regions and in cold regions with sporadic or no permafrost. Some of them have permanent ice cores.

Congelifluction is the name for the downslope movement of material caused by frost action, while *solifluction* is a more general term for downslope movement of material. When the upper part of a frozen soil melts in the spring or summer, it is often saturated with water, and the cohesion between the particles in the wet soil, or the friction at the contact with the frozen ground below, can be so small that the soil moves downslope by gravitational forces. Another important factor is the fact that soil raised by freezing on a slope drops vertically

when the active bed melts, and it therefore drops to a position downslope from its original position. Congelifluction processes are most active in fine-grained material that can hold water, and very little or no congelifluction takes place in coarse material which is well drained. The most common visible congelifluction or solifluction features are solifluction lobes and solifluction terraces and steps.

As previously mentioned, many of the frost features develop to a certain degree in all three of the principal frost zones, while some features are restricted to the permafrost zones, and can be used as climate indicators.

Soil formation (Fig. 3–47)

Chemical and mechanical weathering are dominating factors in soil formation. Most important are the chemical processes. The vegetation and accumulation of organic matter in the upper zone (A_l) produce organic acids which vastly increase the chemical weathering. All of the most soluble minerals, like carbonates and dark minerals, gradually dissolve and

Fig. 3–47. Soils which commonly occur in cool climates. A: Brown soil. B: Chernozem soil (photo by Harald Bergseth). C: Buried podsol (a paleosol). The A_1, A_2, B and C horizons are marked. Only white quartz grains are left in the A_2 horizon. All other minerals are dissolved, and the dissolved matter is deposited in the B horizon.

A

B

C

disappear first from the upper inorganic zone, the A_2 zone. Therefore, the least soluble minerals, like quartz and aluminum minerals, for example, can be enriched in this zone. Water is an essential factor for the weathering process, and in warm moist areas even the quartz may dissolve; the end product can be a red-colored iron and aluminum oxide called laterite soil, which we know well from many warm countries at low latitudes. (See the section on clay minerals in the Glossary.)

Many different types of soils are recognized by specialists, and the soil type and soil development may give some indication about the climate and duration of soil formation. Within the cold Arctic and Subarctic regions the soil formation is slow, and in many regions which were covered by the late Quaternary glaciers, the time for soil formation has been so short that the soils are relatively immature. Figure 3–47 shows three of the traditional main soil types in Subarctic and temperate regions. Buried soils (paleosoils or paleosols) are frequently found in stratigraphic sections interbedded with glacial deposits. They represent soil formation in "warm" interglacial or interstadial periods (Fig. 2–9).

Changes in marine sea level caused by isostatic, eustatic and geoidal changes

Isostasy (Figs. 3–48 to 3–51)

The vertical motion of sections of the earth's lithosphere forced by loading and unloading is called isostatic adjustment. According to the most accepted explanation, the lithosphere, which includes the crust and the upper, rigid part of the mantle (the lithosphere), rests on, or "floats" on, the viscous part of the mantle, the asthenosphere. With increased load on a part of the crust, the lithosphere will sink into the asthenosphere until it reaches a "floating" balance (isostatic equilibrium), and the opposite happens when the load decreases. In addition, with increased load, the "rigid" lithosphere is also elastically compressed, and it expands correspondingly when the load is removed. However, the elastic factor is considered very small compared with the isostatic. The late Cenozoic ice sheets in North America and northern Europe were probably more than 3000 m thick. Therefore, the two northern hemisphere ice sheets represented very heavy

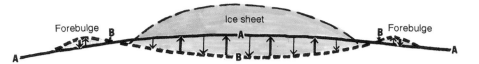

Fig. 3–48. Isostatic depression and uplift of the earth's crust caused by the load and the subsequent unload by melting of an ice sheet. Thin arrows: The isostatic depression below the ice sheet, and the corresponding uplift of the forebulge. Thick arrows: Isostatic uplift of the depressed crust, and the drop of the crust at the forebulge. The magnitude of the depression below the large ice sheets was several hundred meters, and the uplift of the forebulge was a maximum of a few meters.

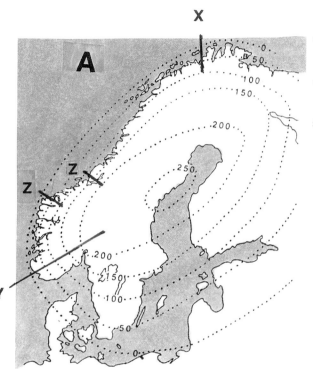

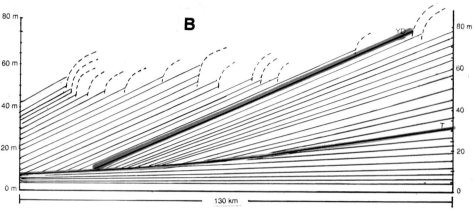

Fig. 3–49. A: The isostatic uplift of Fennoscandia (see Fig. 1–15). x: cross section for the shoreline diagram (Fig. 3–49B). y: cross section for the shoreline in Fig. 3–51B. z: cross sections for the graphs in Fig. 3–51A.

B: Shoreline diagram for western Finmark in northern Norway. (Modified from M. Marthinussen, 1960.) Only the two colored lines (YD and T) are very distinctive, and they were observed by A. Bravais in 1836 (see p. 16). The highest (oldest) lines stop at ice-front positions (dashed, curved lines), which in some cases are marked by end moraines. Line YD is of Younger Dryas age (11 000–10 000 years old), formed beyond the fronts of the glaciers which deposited the large and distinctive Tromsö-Lyngen end moraines. Line (T) is a "broken" line called the Tapes line, which was formed during the Holocene warm period between 8000 and 4000 years ago. The Tapes line consists of segments of several lines. Shoreline diagrams of this kind are constructed for many areas in Fennoscandia, and they are strikingly similar.

weights loaded on the earth's crust, and the lithosphere sank underneath them. Since the specific weight of the ice is about 0.9, and the specific weight of the mantle material is in the order of 3.3, the 3000 m thick ice could theoretically have forced the crust down nearly 1000 m to reach a complete equilibrium (full isostatic compensation). However, this process was slow due to resistance of the rigid lithosphere against being bent down, and due to the slow motion of viscous asthenosphere material from the zone below the ice sheet to the surrounding zones. It was so slow, and the time that the ice-age ice sheets kept their maximum size was so short, that equilibrium corresponding with the 3000 m ice weight was never reached. In fact, a considerable melting

and thinning of the ice must have taken place before the equilibrium was reached. As deglaciation and thinning progressed after this point of balance, the depressed lithosphere started to rise, and is still rising both in Canada and in Scandinavia 8000 to 9000 years after the ice load completely disappeared.

The present-day isostatic rise of the earth's crust in Scandinavia is shown in Fig. 3–52. The maximum rate is approximately 0.8 m per 100 years in the northern part of the Baltic Sea area, where the ice sheet was thickest. The total elastic depression and rise of the crust caused by the ice sheet was most likely in the order of a few meters, and it occurred momentarily, with no time lag, when the ice melted.

The highest isostatically raised shorelines lie

Fig. 3–50. Isostatically raised river terraces near the mouth of Gauldal Valley on the coast of western Norway. The upper terrace represents an outwash delta which was graded to a sea level about 180 m above the present. The present river level is slightly higher than present-day sea level.

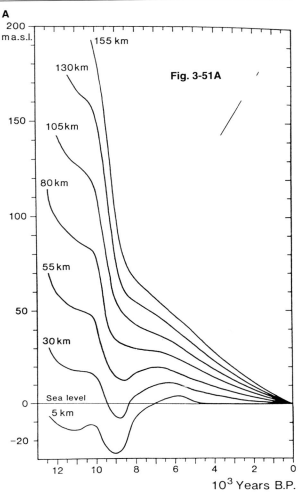

about 300 m above the present sea level both in central Scandinavia and in the Hudson Bay area of North America, where the Quaternary ice sheets were thickest. In general all shorelines rise towards the centers of uplift (see Fig. 3–49), and isobases, which are lines through points with equal uplift, are more or less circular around the same centers (see Figs. 1–15A, 1–16A).

The isostatic depression of the rigid lithosphere, combined with the transport of viscous mantle material from the isostatically depressed glaciated areas to the surrounding areas, caused a small rise of the crust, called the peripheral bulge, in areas at some distance beyond the ice margin (see Fig. 3–48). The process was reversed when the ice sheet melted, and the bulge migrated inwards and dropped.

Eustasy (Fig. 3–53)

Vertical changes in global sea level caused by variations in the volume of water in the ocean basins, variations in volume of the ocean basins, or temperature variations of the ocean waters, are usually called the eustatic variation, and they are of the same magnitude all over the globe.

Variations in water volume

The changes in water stored in the world's ocean basins are undoubtedly the most im-

portant factor for the late Quaternary eustatic changes. Water, temporarily stored in the large ice sheets on the continents, was evaporated out of the world oceans, and ice-volume calculations, together with calculations based on observed submerged and raised shorelines, suggest that the corresponding drop in global sea level was in the order of 100–150 m.

Variations in the volume of the ocean basins

These variations, caused by crustal motion induced by plate tectonics, are undoubtedly important for long-term sea-level changes. However, they were of less importance for the rapid, most-recent late Quaternary changes. But the isostatic depression and rebound of the sea floor caused by the increased load and unload of a 100–150 m water column could have caused some change in the volume of the ocean basins and the associated sea level.

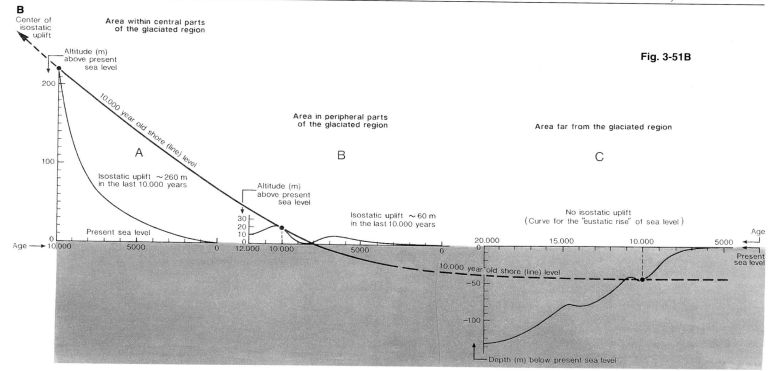

Fig. 3-51B

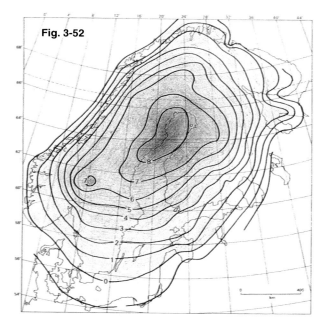

Fig. 3-52

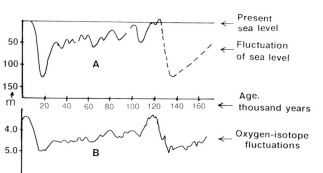

Fig. 3–51A. A suite of shore-displacement graphs from localities along profiles z shown in Fig. 3–49A. The graph marked 5 km represents an area on the outermost coast where the isostatic uplift has been low, and the graph marked 155 km represents an area 150 km further inland where the isostatic uplift has been high. The graphs are based on dated shore levels, and they represent areas which lie 25 km apart (see Fig. 3–49A). (From J. Svendsen and J. Mangerud, 1987.)

Fig. 3–51B. Shore-displacement graphs (A, B, C) and a reconstructed 10 000 year old shore level (line) along a profile (y in Fig. 3–49A) from an area near the center of isostatic uplift in Fennoscandia to the periphery with no isostatic uplift. The illustration shows how shore-displacement graphs can be used to reconstruct old shore levels (lines).

A: Shore-displacement graph for the Oslo region (modified from U. Hafsten, 1958). B: Shore displacement in southwestern Norway (modified from U. Hafsten, 1983). C: A very generalized graph which presents the approximate "global" eustatic rise of sea level. Most of the previously published graphs lie fairly close to this one. However, the graphs for the observed "eustatic changes" are different in different parts of the globe, due to geoidal changes.

The dashed part of the 10 000 year old shore level (line) represents a theoretical shoreline in the North Sea region. However, this region experienced tectonic crustal movements, and the dashed shore level (line) may not correspond with the true 10 000 year old shoreline. The constructed shoreline is bent, but keep in mind that the distance from A to C is extremely long compared with the measured altitudes.

Fig. 3–52. The earth's crust below Fennoscandia is still rising as a result of the unloading of the heavy Weichselian Ice Sheet. The map shows the present-day annual isostatic uplift in mm. The maximum uplift (in the center) is more than 8 mm per year, or more than 80 cm per 100 years, and note that the ice sheet disappeared about 8500 years ago. (From J. Kakkuri, 1991.)

Fig. 3–53. Fluctuations of the global sea level (the eustatic fluctuations) during the last 150 000 years, combined with a deep-sea oxygen-isotope graph. (Modified from J. Chappell and N. Shackleton, 1986.)
Graph A probably shows a relatively good approximation of the fluctuations of the global sea level. However, local geoidal changes have affected sea-level fluctuations in all parts of the globe, and no single graph exists which is valid for the entire globe.

Geoidal changes

The transport of large water volumes from the ocean basins to form the huge ice sheets on the continents during the ice age resulted in changes in the shape of the global geoid. Calculations of these changes have been done, and they indicate that the geoidal change in sea level could have been considerable in some areas. Unfortunately, the ice age shore levels around most of the world's continents lay well below present sea level, and therefore direct observations of the ice age shore features, which could lead to good reconstructions of the geoid, are almost absent. The geoidal changes in sea level are of course very difficult to separate from the true eustatic changes. Therefore, the geoidal component, called geoidal eustasy, is usually integrated in the observed eustatic changes.

Graphs of the eustatic changes

It was previously believed that the change in water volume was the only important factor for at least the rapid eustatic changes during the last 20 000 years, and it was also believed that the eustatic changes were the same along all coasts. Therefore, several scientists attempted to reconstruct "the general graph" for the eustatic changes on the basis of both shore-level observations and theoretical ice/water-volume calculations. However, it is impossible to determine a eustatic graph which is valid for the entire globe, due to the geoidal effect. Most graphs are integrations combining the true eustatic and the geoidal changes, and they are significantly different for different parts of the world. However, in general the reconstructed graphs for various regions are very similar, and they show a general trend of global eustatic changes (see Fig. 3–53).

Vertical shoreline displacement (Fig. 3–51A)

The combined effects of isostatic, eustatic, and geoidal changes determine the altitude of the shore level at any place and at any time, if the area was not subjected to tectonic changes. In tectonically unstable regions like California

and Italy, the tectonic effect was of primary importance, and some tectonically elevated Quaternary shorelines lie several hundred meters above present sea level. In most of eastern North America and northern Europe, the tectonic factor seems to have been small. Within the glaciated regions, where the isostatic factor was large, the result of the combined isostatic, eustatic, and geoidal changes has been presented by shoreline-displacement graphs, generally constructed on the basis of pollen and diatom analysis of the sediments in radiocarbon-dated cores from lakes at different elevations on the coast. The age of the transition from marine to lacustrine conditions in the lake basins must be determined together with the exact altitudes of the lake thresholds. These facts provide the basis for the construction of the shoreline-displacement graphs. In areas outside of the glaciated regions, the ice age and early Holocene shorelines are usually submerged below present-day sea level, and the shoreline displacement is difficult to determine. Therefore, only a few graphs have been constructed in these kinds of areas, and they generally correspond approximately with the graphs for the general eustatic rise of sea level (see Fig. 3–53).

FIELD AND LABORATORY METHODS

The previous chapters were focussed on some processes which must be understood to be able to reconstruct a Quaternary history. The following discusses some of the field and laboratory methods which were used to determine the origin and age of the sediments, and to reconstruct the environments in which they were deposited. The fossil content, the chemical properties, and the lithology of the sediments are all important in this connection, and many different scientific specialists are involved in studying them. In addition, the economic aspects and the land-conservation aspects of the Quaternary sediments must be dealt with, since they represent a most important part of the Quaternary science. Several books have been written about each of the above subjects, and students who want a deeper knowledge are referred to them.

Stratigraphy

Stratigraphic studies focus on the sedimentary beds in stratigraphic sections in which information about the geologic history is stored. The fossils, the lithology and various other sedimentary properties can all be studied, and several different methods are used to date the sediments. Traditionally there have been three major stratigraphic branches: lithostratigraphy, biostratigraphy and chronostratigraphy. However, several new branches have been introduced recently, mainly as a result of rapidly expanding scientific achievements. Isotope-stratigraphy and paleomagnetic stratigraphy are good examples (see Glossary, p. 190). The following is a brief review of the

stratigraphic branches and some principles on which stratigraphy is based.

Lithostratigraphy

The size, shape, and orientation of the grains in the sediment, the sedimentary structures like various types of stratification and tectonic features, and the geotechnical properties and the petrography/mineralogy of the clasts are the most important aspects for lithostratigraphic studies (Glossary, p. 189).

In describing sediments within a bed, or unit of several beds, descriptive terms should be used. For instance, terms like diamictons, gravel and clay are preferable to genetic terms such as till, glaciofluvial gravel, and marine clay, which involve not only a sediment description but also a genetic interpretation. These terms should be used only when the descriptive parameters have been evaluated

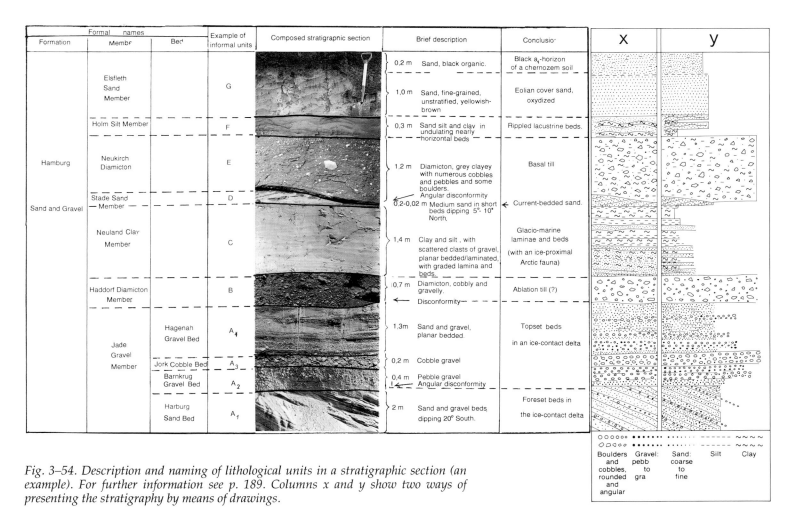

Fig. 3–54. Description and naming of lithological units in a stratigraphic section (an example). For further information see p. 189. Columns x and y show two ways of presenting the stratigraphy by means of drawings.

and a conclusion on genesis has been drawn.

The sediment bed (unit) is generally named after the dominant grain-size fraction, and names of other important fractions may be added as adjectives. For instance: silty sand; clayey-silty sand. Sediments that are diamictons or contain more than about 20% clay are usually called diamictons and clays, respectively, and the names of other important fractions are added as adjectives. For instance: sandy diamicton; silty clay. Additional descriptive terms may be added after the sediment name, such as: silty clay, stratified with scattered pebbles. In some cases both adjectives and other descriptive terms are placed after the primary name, for instance: sand, silty with scattered pebbles. This is done particularly in connection with descriptions added to stratigraphic columns (see Fig. 3–54).

Variations in grain size and sorting, for example, are generally used to distinguish the different lithostratigraphic units, and a unit can be subdivided into subunits. For instance, a sand unit may contain some silt or clay beds, which may call for a subdivision of the unit. Figure 3–54 shows an example of this. According to the international lithostratigraphic code, a geographic name together with a grain-size fraction and a unit name should be used to formally name the unit. For example, Boston Sand Formation, Michigan Silt Member, and Valders Clay Bed. However, there are certain restrictions in using the formal nomenclature, and therefore a more informal naming of the lithostratigraphic sediment units is more frequently used (see Fig. 3–54 and Stratigraphy in the Glossary).

Biostratigraphy

The fossil content of the beds and the information the fossils render about the organic life and environment in which the plants/animals lived are the main objects for biostratigraphic studies. Biostratigraphic units are called "zones". The fossil content within a zone is relatively uniform, and it is different from the fossil content in the zones above and below. Names of the most dominant species or the most important species are used to identify the zone. For instance, the *Betula Zone*, the *Betula-*

Corylus Zone, the *Mytilus Zone*. Note that the biostratigraphic zones do not necessarily correspond with the lithostratigraphic zones (see Fig. 3–55). Conclusions about the climate and other environmental factors can be drawn by comparing the fossil assemblage with the present-day distribution of the corresponding plants/animals. Studies of the different fossil groups are now so advanced that scientists frequently specialize within one single plant/animal group. In this book only brief examples of some of the most important groups will be mentioned. More detailed information is presented in several textbooks, and some information is presented under Stratigraphy in the Glossary.

Macro-fossils

Important groups of macro-fossils are plant/tree remains, vertebrates and invertebrates (molluscs, snails and beetles, for example). They represent a significant component of Quaternary stratigraphy.

Micro-fossils

Pollen. In terrestrial sediments the pollens from trees and plants are the most commonly used fossils in distinguishing the vegetation zones. Millions of microscopic pollen grains are spread mainly with the wind, and come to rest in lake and bog sediments. Changes in vegetation through time can be determined by counting the pollen grains from the different plants and trees in cores from bogs or lakes, allowing for differences in sedimentation rates and differences in pollen production among the various species. The radiocarbon method is generally used to date the pollen zones.

The result of the pollen analysis of a stratigraphic section (a core) is generally presented as a pollen diagram (see Figs. 2–12, 2–68). On the basis of changes in the vegetation, the different vegetation zones are distinguished, and they are named as indicated above. However, for late-glacial and postglacial deposits in Europe, an old, traditional numbering of zones from I (the oldest) to VIII (the youngest) is still in use (see Fig. 2–68 and Glossary).

Pollen zonation has been applied in correlating and dating the deposits in stratigraphic sections, and the numbered pollen zones (from I to VIII) are generally used as reference zones. However, numerous radiocarbon dates of the zones show that their ages vary slightly from one area to the other. This has to do with the migration pattern of the vegetation. Most zones are time-transgressive, and the boundaries of postglacial zones in Scandinavia are frequently slightly younger than corresponding boundaries in Germany, for instance.

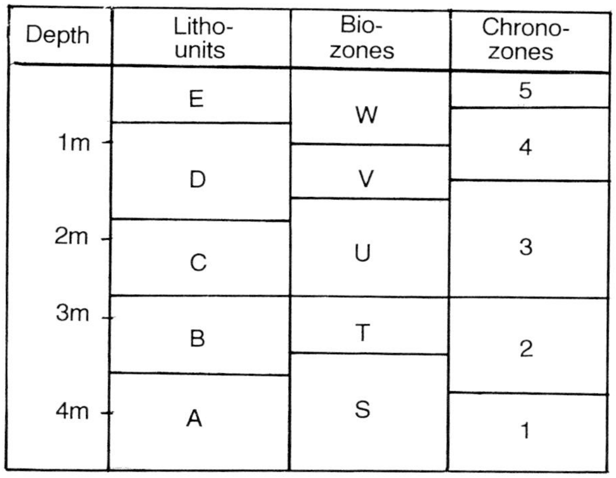

Marine micro-fossils

As mentioned before, the new drilling techniques and ships with advanced drilling equipment have opened a new world, the submarine world, for stratigraphic studies. This has stimulated research in various fields, including marine micro-fossils. Numerous long cores from the floors of all the world's oceans are now being studied, with remarkable results. The micro-fossils found in the cores are either of planktonic species (which drifted with the surface-water currents) or of benthonic species (which lived on the sea floor). The planktonic forms are both primitive plants and primitive animals which serve as food for higher forms of organisms. They record conditions in the surface water of the oceans. The following groups are those most used in biostratigraphic studies: *Foraminifera* (both benthonic and planktonic species), *Ostracods, Coccoliths, Radiolaria, Diatoms*, and *Dinoflagellates*. Some species of diatoms live in brackish water and some live in fresh water. Therefore, diatom stratigraphy is commonly used to record the transition from a marine to a fresh-water environment (see p. 144).

Chronostratigraphy

As already mentioned, the boundaries between lithostratigraphic or biostratigraphic units are usually time-transgressive. For instance, the age of the lower boundary of a till bed or a bio-zone in one area can be older than the corresponding boundary in another area. This led to some confusion, and the need for zones with time-parallel boundaries developed. These zones have usually been called

Depth	Litho-units	Bio-zones	Chrono-zones
1m	E	W	5
			4
	D	V	
2m	C	U	3
3m	B	T	2
4m	A	S	
			1

Fig. 3–55. Chronozones, bio-zones and lithological units in a 4.5 m stratigraphic section. Note that the boundaries between the chronozones do not necessarily correspond with the boundaries of the bio-zones and the lithological units.

chronozones. Figure 3–55 presents a chart which indicates the relationship between lithological units, bio-zones, and chronozones in a stratigraphic section (see also the Glossary, p. 190).

The ability to date the zone boundaries depends upon the material available for dating, the quality of the material, and the dating method which can be used. Before the discovery of radiometric dating methods, only a few exact dating methods existed, such as the varve method (varve chronology) and the tree-ring method (dendrochronology), but their use was very limited. The introduction of the radiometric method vastly increased the possibility of "exact" dating. It also opened the possibility to calibrate several other non-radiometric methods, and this led to the discovery of several new methods, which will be described in the following sections.

Radiometric dating

Radioactive isotopes disintegrate and radiate particles at fixed rates. The rates are indicated by the so-called half-life of the isotopes, which is the time it takes to disintegrate one-half of the original amount of an isotope. As a result, the radioactive emission of particles is also recorded in terms of half of the original, and can be measured by modern instruments. Therefore, the half-life of an isotope establishes the rate of the "geologic clock" that the isotope represents. The half-life also gives an indication for which geological time bracket the clock can be used.

The radiocarbon method

The method is based on the assumption that the amount of radiocarbon (^{14}C) in the atmosphere, and hence in living organic matter, has been constant during the last 70 000 years. After the time of death of the organism there is no further input of carbon, the "clock" is started, and the ratio of ^{14}C/^{12}C shifts as the radioactive ^{14}C disintegrates. Therefore, measuring the ratio provides an estimate of elapsed time since the death of the organism. Today we know that the basic assumption is not completely correct (see Glossary). An important part of a field scientist's work deals with finding and collecting, from the proper context, organic matter which can be radiocarbon dated, and providing the correct stratigraphic setting and interpretation. Dating is carried out in the laboratory, usually by one of the two following methods:

1. The traditional gas-counting method, in which the emission of radioactive particles is measured.
2. The AMS-method, where accelerator mass spectrometers are used to measure directly the mass-ratio between ^{14}C and ^{12}C.

Most laboratories cannot adequately date material older than 40 000–50 000 years, but dates up to 70 000 years old have been presented. Most scientists now consider the high-age dates to be minimum ages (see Glossary).

The Uranium-series method

This method is based on measuring the ratios between the radioactive uranium and thorium isotopes which are found in corals, speleothems (calcite layers, stalagtites, and stalagmites), marine shells, and peat. Material from about 1000 to about 350 000 years old can in general be dated by this method, and still considerably older samples have been dated by means of improved higher-precision methods. Knowledge about the isotopic ratio at the zero point, when the sediment was deposited, is needed to obtain correct results, and in some cases this is problematic. The method seems to give the best result for corals and calcite deposits (see p. 183).

The potassium-argon method and the argon-argon method

These are used to date the oldest Quaternary sediments, generally older than 500 000 – 1 million years (see p. 185).

The thermoluminescence (TL) method

Minerals like quartz and felspar, which lie in a sediment, experience a gradual change of their atomic structure caused by the radioactive radiation which the minerals are exposed to within the sediment. This change takes place at a rate which depends upon the intensity of the radiation, and that can be measured with counters (dosimeters) placed in the sediment in the field, or in a sediment sample brought to the laboratory. The accumulated change within the minerals can be measured by heating the minerals and registering the light generated as the atomic structure readjusts to its original state. In that way the age can be determined. "The start of the clock" took place when the sediment was deposited and the minerals were last exposed to the sunlight. This start is recorded as a "plateau level" in the registered graph, and it is important that this level is clearly marked. For sediments which were not exposed to much light when they were deposited, the plateau level is generally weak. Therefore, the method appears to work well for some sediments, and not for others.

The thermoluminescence method has been extensively used to date prehistoric pottery, and in such a case the zero-setting of the clock took place when the clay was heated to make the pot.

The optical stimulated luminescence (OSL) method

This is based on the same principles as the TL method, except that optical light is used instead of heating to stimulate the measured light from the minerals.

ESR dating (electron-spin resonance dating)

Paramagnetic defects of the crystal structure are created by the ionizing radiation of radioactive elements. The defects can be detected by ESR spectroscopy, and the intensity of the ESR signal corresponds with the annual dose-rate and the length of time for which the crystal has been exposed to the dose. Promising ESR datings have been carried out on marine shells, and it is indicated that the method may be used on spleothems, bones and various appropriate sediments. The dating range will probably cover much of the Quaternary Period when the technique is refined.

Surface-exposure dating

This is a promising new method which is now being tested. It is based on measurements of cosmic-ray-produced nuclides (^{36}Cl, ^{3}He, ^{10}Be, ^{21}Ne, ^{26}Al) on exposed rock surfaces. However, only minimum ages are obtained by this method if the rock surface was eroded or covered with sediments during parts of the time interval which we want to date.

The fission-track method

This method is based on counting the "tracks" which emission from fission of radioactive isotopes (^{238}U etc.) produce on the surrounding crystals in volcanic deposits. Again, the intensity of the radioactive emission in the deposit must be measured to carry out age calculations.

Other new important dating methods

Many of the new important dating methods are based on processes which proceed at more or less fixed rates, and radiometric dates have been used to calibrate the rate/time scales for most of them.

The amino-acid method

This is based on recording the alteration (called racemization or ephimerization) of the proteins which takes place in dead organisms. At a constant temperature this alteration takes place at a fixed rate, but unfortunately the rate is very temperature-dependent. The method has been much used for approximate age determinations, and used with care it can be very important, in particular in determining the relative age of sediments in one section or adjacent sections.

The paleomagnetic methods

Paleomagnetic reversals and secular changes. Lacustrine and marine sediments, loess, volcanic ash and lava etc. generally contain magnetic minerals which were oriented in the global magnetic field when they were deposited. This original orientation can be observed today by means of sensitive instruments, and in that way the direction of the former global magnetic lines, the paleodirection, can be reconstructed. Since the magnetic poles are constantly moving, and have moved considerably during the earth's history, the observed paleodirection may vary considerably from one bed to the other in a stratigraphic section, and the changes may be so characteristic that they can be identified and dated by comparison with previously dated, standard paleomagnetic reference diagrams.

In fact, there are two paleomagnetic methods, one in which the relatively rapid and drastic changes – the paleomagnetic reversals – are recorded, and another which focusses on the smaller, gradual secular changes. During a paleomagnetic reversal the global magnetic field is reversed, and the magnetic North Pole becomes the South Pole. The pattern for the paleomagnetic reversals is shown in Fig. 1–19. The secular changes have so far been used mainly to date relatively young sediments, of Late Weichselian and Holocene age.

The oxygen-isotope method

This method, when coupled with radiometric dating methods, is now generally used to date deep-sea deposits. The oxygen-isotope pattern in deep-sea deposits, which records changes in global glacier volume, is so characteristic, well established, and well dated that standard reference diagrams are now used to date deep-sea deposits from all of the world's oceans (see p. 181 and Fig. 1–19).

Fig. 3–56. Cross section of a tree with 20 annual rings. The dark rings are winter rings. (From an exhibition at The American Museum of Natural History, New York.)

Fig. 3–57. A varve sequence from central Sweden. The dark-grey clay layers represent winter deposits, and the thicker buff-colored layers with silty clay represent summer deposits. The photo to the right shows the 17 varves of unit x. (Photo by Lars Brunnberg.)

grams, a chronology which goes back to 12 000–13 000 years ago has been established in Sweden.

Lichenometry

Lichens which grow on rocks increase in size (area) with age, and some lichen species have been used to date very young end moraines on the basis of the assumption that the lichen started to grow on selected boulders on the moraine surfaces just after the moraines were deposited.

Biostratigraphic methods

Biostratigraphic zones (pollens and forams, for example) are generally time-transgressive, i.e. the ages of the zone boundaries change from one area to the next. However, if the zone boundaries are well dated in one stratigraphic section, this section can generally be used as a reference section to date the corresponding zone boundaries in adjacent sections.

Relative-dating methods

The "dating" of bio-zones and litho-zones by correlating zones in one area with zones in another is called relative dating, but the method cannot be used in exact dating of zones over considerable distances. However, over very short distances, the correlation with well-dated zones can be used in more exact dating, as indicated above.

Traditional dating methods

Dendrochronology

This is based on counting annual tree rings and plotting their thicknesses in diagrams. Tree rings of both living and dead trees are used (Fig. 3–56). Dead trees as old as 10 000–12 000 years have been dated by correlating and matching tree-ring thickness graphs.

Varve chronology (Fig. 3–57)

This was developed by G. De Geer in Sweden. He counted and measured thicknesses of varves, annual clay/silty couplet layers, deposited in the former Baltic Sea, which covered much of Sweden. The method is in principle the same as that in dendrochronology. By matching and overlapping varve-plot dia-

The physical-mineralogical character of the sediments

As previously mentioned, the environment in which the sediments were deposited is reflected in their physical-mineralogical properties, and these kinds of properties determine various practical economic applications of the sediments. Therefore, they are essential aspects of Quaternary geology. However, it must be kept in mind that the information on these subjects is so extensive that only a brief introduction to some aspects is possible.

Sedimentary grain sizes

The grain size and the grain-size distribution of a sediment are two of the most important and useful physical characteristics of the sediment. They are frequently semi-diagnostic in interpreting the origin of the sediment, and they are highly important for much practical use of the sediment.

Grain-size classification

Several different classification systems have been presented in other literature, but the "old" Atterberg system and the more modern Wentworth system are used the most. The two systems are presented under grain size in the Glossary. The Wentworth system is now generally accepted, and Wentworth fractions will be applied in this book. Note that the grain-size scales are logarithmic, and the Phi-scale, which was developed in order to simplify numerical grain-size calculation, is the basis for the Wentworth classification (see Glossary). In Fennoscandia the Atterberg Scale is still frequently used. Observe that this scale shows increasing grain sizes towards the right on the diagram, which is the opposite of the Wentworth Scale (see p. 172).

Grain-size distribution

Grain-size measurements (see presentation in the Glossary). The weight percentages of the different grain-size fractions in a sediment can be measured in several different ways. Fractions coarser than 0.063 mm (coarser than silt) are usually separated by *sieving*. The fine-grained fractions (silt and clay) are measured by methods based on the velocity at which the particles drop in a fluid. Several methods have been developed for this kind of measurement. A simple, much-used, but slow, method is the *Pipette method*. Other, fast methods are based on more advanced instruments. Some of them are connected with computer-driven printers, which print the cumulative graphs directly.

Presentation of the grain-size distribution (Fig. 3–58)

There are many ways in which the grain-size distribution can be presented. Some of them are as follows:

1. *A linear percentage diagram.* The method is demonstrated in Fig. 3–58, and it is frequently used to present the grain-size distribution in stratigraphic sections.
2. *Histograms.* These are generally used to present the grain-size distribution of one sample. Each column represents the percentage of one grain fraction of the sample.
3. *Triangle diagram.* This is used to show the percentages of three fractions of a sample, for instance, sand, silt, and clay. The composition of the sample is presented with one dot within the triangle, and one triangle can be used to present the composition of many samples.
4. *Cumulative graphs.* These give the most complete picture of the grain-size distribution of a sample. The graphs show the weight percentage of all grain-size fractions. There are two kinds of cumulative graphs, and they differ in their use of vertical percentage scale: *The normal diagram* has a normal vertical percentage scale, and *the log-probability diagram* has a log-probability vertical percentage scale. This scale is squeezed together in the middle (near 50%) and stretched towards the 100% and 0% ends. The advantage in using the log-probability diagram is that a log-normal grain-size distribution is presented as a straight line, and a sediment with one grain population usually has this kind of distribution. The graphs for sediments with several grain populations are usually composed of several fairly straight lines (see Fig. 3–58).

Grain-size parameters

Grain-size parameters numerically present certain important properties of the grain-size distribution and features of the cumulative graph. The four most commonly used parameters are median diameter, sorting, skewness and kurtosis. They are all described in detail on p. 174 in the Glossary.

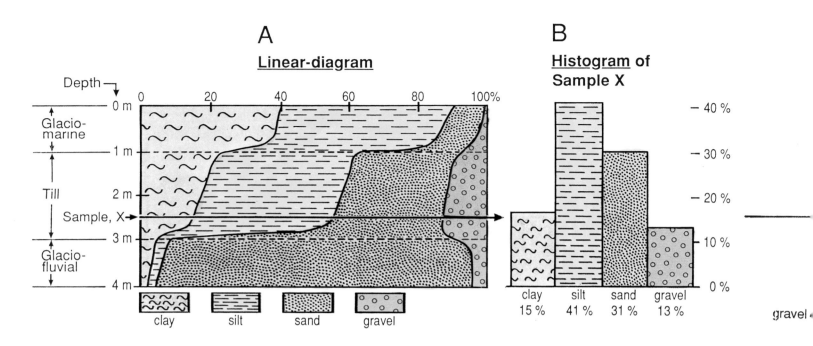

A
Linear-diagram

B
Histogram of Sample X

clay 15 %   silt 41 %   sand 31 %   gravel 13 %

Fig. 3–58. Different ways of presenting the grain-size distribution of a sediment.

A: Linear presentation of the sediment (grain-size) distribution in a 4 m stratigraphic section which consists of glaciofluvial sand (4–3 m) clayey, silty till (3–1 m), and glaciomarine silty clay (1–0 m).

B: Histogram of the grain-size fractions in sample x at 2.5 m depth.

C: Triangle diagram where sample x is plotted as a large dot. All numbers are percentages. Note that only three grain fractions can be presented in a triangle diagram. Samples containing 100% sand, gravel, or silt and clay plot on the respective corners.

D: Grain-size distribution of sample x plotted as cumulative graphs. Graph A_1 is plotted on a log-normal paper where the dashed lines and the normal percentage scale to the right are used. Graph A_2 is plotted on a log-probability paper where the probability percentage scale to the left is used. Sample x is a till, and the many straight lines in A_2 suggest that the till consists of many grain fractions which the glacier picked up from clay, silt, sand and gravel beds which it overrode. Note that the log-probability graph and scale is extended above and below the diagram paper to reach closer to the 100% (99.9%) and the 0% (0.1%) values. This was done in constructing all of the diagrams in this book, even though the parts above the 99% line and below the 1% line are not presented.

The mean diameter corresponds roughly with the grain size at the 50% level on the cumulative curve. The sorting parameter expresses the degree of spread of grain sizes in a sediment. A sediment with grains concentrated near one size is well sorted, and the corresponding cumulative curve is steep. Figure 4–11 on p. 175 presents a sorting scale based on the calculated sorting values. The skewness expresses the degree of departure from a grain-size distribution which is symmetrical on both sides of the 50% level and has a skewness which equals zero. Increasing posi-

tive skewness values express an increasing amount of fine-grained material in a fine-grained "tail" relative to the coarse-grained "tail". Increasing negative values express the opposite: increasing coarse-grained material in the coarse-grained "tail". The kurtosis is an expression for the concentration of grains near the median diameter compared with the amount near the "tails" (see p. 174).

Interpreting the results of grain-size analyses

The grain-size distribution may give information about the genesis of the sediment, its source, its mode of transport, and the environment in which it was deposited. In some cases the distribution pattern can be diagnostic, but in most cases it leaves openings for several alternative interpretations; see p. 172 in the Glossary.

Roundness and shape of grains

The roundness and shape of the grains can be diagnostic in interpreting the mode and distance of transport.

C
Triangle diagram

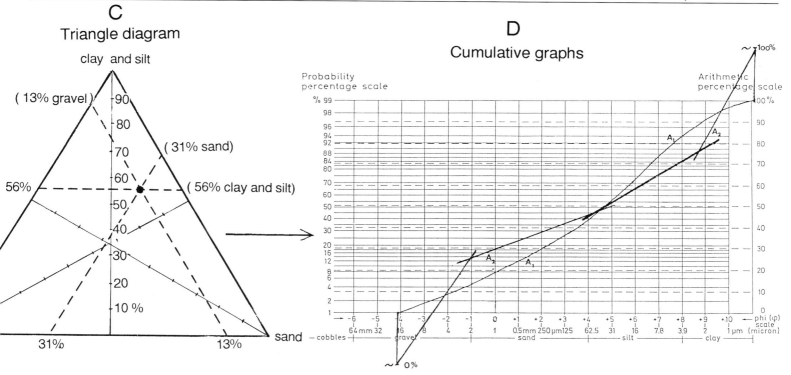

clay and silt

(13% gravel)

(31% sand)

(56% clay and silt)

90
80
70
60
56%
50
40
30
20
10 %

31%   13%   sand

D
Cumulative graphs

Probability
percentage scale

Arithmetic
percentage scale

~ 100%

% 99
98
96
94
92
88
84
80
70
60
50
40
30
20
16
12
8
6
4
2
1

~ 0%

-6 -5 -4 -3 -2 -1 0 +1 +2 +3 +4 +5 +6 +7 +8 +9 +10 ← phi (φ) scale
64mm 32 16 8 4 2 1 0.5mm 250µm125 62.5 31 16 7.8 3.9 2 1µm (micron)

— cobbles — | gravel | — sand — | — silt — | — clay —

A_1 A_2

Roundness. The classification of the roundness of clasts can be based on systems with 4, 5 or 6 roundness classes. In the simplest system, the 4-class, each class corresponds with a major genetic mode of formation:

1. the angular clasts represent broken or crushed rocks;
2. the clasts with rounded edges are typical for glacially transported material;
3. the rounded clasts are typical for short river transport, which is common for many glaciofluvial sediments;
4. the well-rounded clasts are characteristic for sediments with a somewhat longer river transport, or sediments in beaches which have been exposed to wave abrasion for a longer period.

The other systems, with 5 and 6 classes, are more advanced and detailed; in particular, the system with 5 classes is much used today (see Fig. 3–59).

The roundness analyses of clasts in till usually show a varying amount of angular, rounded and well-rounded clasts in addition to the main glacial population, clasts with rounded edges. In such cases the rounded and well-rounded clasts represent a population of former river or beach clasts. The angular clasts usually represent glacially crushed clasts. Crushed clasts, and dented clasts where the

Fig. 3–59. Roundness classification of clasts. A: a 6-class system. B: a 5-class system (1: well rounded, 2: rounded, 3: sub-rounded, 4: subangular, 5: angular).

In a 4-class system the clasts in group 4 (B) are more or less added to group 3 (B).

dents have relatively sharp edges, are fairly common in many subglacial tills. A classification system has been developed in which dented and crushed clasts are included.

The roundness of the stream-transported clasts depends both on transport distance and on grain size. Larger grains (up to medium-sized boulders) are rounded more quickly than smaller grains. Questions have even been raised about whether sand-sized and smaller grains become significantly more rounded with increasing transport distance.

Eolian sand grains usually need to be transported a considerable distance or time to obtain a fair degree of roundness. Very young, short-transported sand populations may contain grains of soft rocks or shells which are better rounded than the hard quartz grains.

Shape. Several systems have been developed in which the shape of the grains is measured and described. They are more elaborate and time-consuming methods, but may provide more detailed information than roundness analyses.

Surface morphology of grains

The detailed surface morphology of the sediment grains also frequently reflects the mode of transport. For instance, glacially transported clasts may be striated and have small chatter-marks. Eolian sand grains commonly have frosted (sand-blasted) surfaces. In particular, studies of small grains by means of the electron microscope have in many cases rendered interesting results, where it has been possible to identify features characteristic of glacial transport, river transport, and eolian transport.

The orientation of clasts

Fabric analysis

The orientation of clasts in a sediment depends upon transport medium, transport direction and depositional environment. The orientation analysis, which is called fabric analysis, has been used in particular to determine the transport direction of the sediments. For details about the method, see the Glossary (p. 166).

Till fabrics

Clasts in *lodgement till* frequently have a significant preferred orientation with the long axis (a-axis) parallel with the transport direction, the glacier-flow direction. The a-axis usually plunges slightly opposite to the transport direction, and flat stones commonly have a preferred dip in the same direction – they are imbricated. The orientation of the clasts in *melt-out* (ablation) tills often varies considerably. Tills that melted out from the base of the glacier may have a significant preferred clast orientation parallel with the glacier-flow direction because the orientation was established while the clasts were located in the glacier sole. However, tills melted out on the glacier surface frequently have no preferred clast orientation, or they may have a random or a slightly preferred orientation in any direction, including parallel with the glacier-flow direction. The long axes of clasts in *flow tills* are usually orientated essentially parallel with the flow direction, which is not necessarily parallel with the flow direction of the glacier.

Fabrics in river deposits, mud flow and beach deposits

The a-axes (long axes) of stones on the *river beds* frequently have a preferred orientation parallel with the transport direction. This is particularly true for turbulent rivers where the stones are bounced into different positions, and stones which happen to orient with the a-axis parallel with the flow direction rest more stably than other stones. Therefore, a selection and enrichment of stones with this kind of orientation takes place. On some river beds, where the transport is less turbulent but the water discharge is variable, the stones may receive a preferred transverse a-axes orientation, which is the normal orientation of stones which roll on the river bed. However, river-transported stones, in particular the flat stones, are generally imbricated with an upstream dip of the flat surfaces.

The a-axes of clasts in *mudflow, slope wash,* and *slope creep deposits* commonly receive an orientation parallel with the transport direction, i.e. downslope. Stones in *beach ridges* are frequently oriented with the a-axes parallel with the ridge, and dropstones in *glaciomarine deposits* generally have no clearly preferred orientation direction.

Clast-supported and matrix-supported sediments

Coarse grains in a sediment with a finer-grained matrix, may lie in direct contact with each other and support the sediment structure, or they may "float in the matrix", predominantly without this contact. The first type of sediment is clast-supported, and the second type is matrix-supported. River deposits, beach deposits and eolian deposits are generally clast-supported, and the coarse clasts in tills, mudflows and slump/creep deposits are usually matrix-supported.

Tectonic structures

Different kinds of glaciotectonic structures in sediments that were overridden by the Quaternary glaciers have been used to determine the former ice-flow directions (see p. 192).

Petrology and mineralogy

Stone counts, indicator boulders and boulder trains

The petrology of the coarse-grained fraction in a sediment can be determined by means of so-called stone (pebble) counts. Generally the petrology of 100 stones is determined, and the percentage of each petrographic fraction is calculated. If accurate bedrock maps are available, the distribution of the source rock for each fraction can be observed, and in that way the transport direction of the sediment may in many cases be determined. However, frequently the source rock has a wide distribution, and no clear transport direction can be observed. Some tills contain characteristic erratics from very limited source-rock areas, the so-called indicator erratics, which may have a fan-shaped or linear distribution on the lee-side (downglacier side) of the indicator-rock outcrop. They form a "boulder train" or "boulder trail" (see Figs. 2–5, 2–6).

In some cases stone counts and boulder trains (trails) have been used in searches for ore deposits, and stone (pebble) counts are important in determining the quality of the gravel for various practical purposes.

Mineralogy

The mineralogy of the clasts in a sediment is usually determined by means of magnifying glasses or microscopes, and heavy liquids are used to separate the heavy minerals. Of particular interest is the mineral composition of the clay fraction, which is determined by means of XRD analysis (see p. 194). The clay-mineral composition often renders information about the source rock of the sediment, as well as the degree of weathering and diagenesis of the sediment (see p. 162).

Calculating the carbon content (the loss of ignition)

The content of *total carbon* (TC) in a sample is determined by heating a small portion of the sample and burning away the carbon. This is generally done in furnaces which automatically register the produced CO_2 and calculate the weight percentage of carbon. The content of *total organic carbon* (TOC) in a sample is calculated by adding heated HCl until all carbon in the carbonates has disappeared as CO_2, and the amount of organic carbon is calculated by "burning" the remaining part of the sample as indicated above.

All of the carbon in TOC is a result of organic production, and in marine sediments from many ocean areas, the TC reflects the production of carbonate shells. Therefore, both TC and TOC have been used as important stratigraphic tools. TC and TOC graphs of stratigraphic sections, for example, marine cores, frequently correspond well with observed bio-zones and isotope zones.

Mapping of surficial deposits

The youngest geological history is usually recorded in deposits exposed at or near the surface of the terrain. The surficial deposits reveal, in general, more details than the older deposits, and they are the object of broad-scale mapping carried out in most countries. This mapping includes mapping of the morphological surface features, and it is generally combined with stratigraphic studies of the uppermost young beds below the surficial beds.

In addition to recording the younger geological history, the surface deposits are important for much human activity, such as farming, forestry, and various other industries, as well as water supplies. Considering the importance of the surface deposits, all good land-use planning and land conservation must be fundamentally based on their geological characteristics. In that connection the geological maps of the surficial deposits are basic and essential.

How is the mapping carried out?

The mapping of the surficial deposits includes observation of their geographic distribution, grain sizes, sediment textures and structures, and geomorphological features, as well as interpretation of the genetic history of the deposits and features. In addition, the sediments must be sampled for more accurate grain-size analyses, rounding analyses, petrographic analyses, mineralogical analyses (clay minerals), geotecnical analyses, studies of fossil content, and age determinations.

The type of mapping which is carried out and the kind of maps that are produced depend upon the purpose of the mapping, and upon the kind of deposits that are present within the study area. There is no globally accepted code or nomenclature for the symbols and colors used on the Quaternary maps. However, by tradition the symbols are much the same in various countries, and they are always explained in the map code. Therefore, trained workers have generally no problems in reading the maps. In addition, the maps are usually accompanied by a text, where features are explained.

There is a limit as to how much information can be presented on one map. For this and other reasons thematic maps are frequently produced. They present selected themes, such as the distribution of end moraines (as in Figs. 2–47, 2–48) or different kinds of tills; marine deposits; loess; sediments of different grain sizes (clay, silt, sand); and economically important gravel and sand deposits.

General inventory maps which are produced by the geological surveys in most countries include information of general interest. They usually record the deposits in about the upper 1 m of the ground, but they are frequently accompanied by stratigraphical profiles which present information about the subsurface sediments and stratigraphy also. Various symbols are used for the different grain sizes and for the geomorphological features. Colors are generally used for the various genetic sediments, such as tills and glaciofluvial deposits. In areas with surficial deposits from different ice ages, the colors may be used to distinguish deposits of different ages. Figure 3–60 is an example of a general map of surficial deposits made by the Norwegian Geological Survey.

Economic aspects

There are many important economic aspects related to Quaternary deposits. A few of them will be mentioned briefly:

1. *Farming and forestry*. These generally depend upon the quality and distribution of Quaternary deposits. Some of the best farming districts in the world lie on glacio-lacustrine and glacio-eolian deposits, while large areas covered with tills, glaciofluvial/fluvial and glaciomarine/marine deposits represent excellent forest land or excellent-to-poorer farmland. The quality of the sediments for farming and forestry depends to a considerable degree on the source rocks for the sediments.

2. *Industries based on Quaternary sediments*. Many different kinds of industries utilize Quaternary sediments. One of them, the mining of sand and gravel, is economically one of the most important mining industries in the world. Other important industries use sand, gravel, and clay in various products for house construction, and peat is

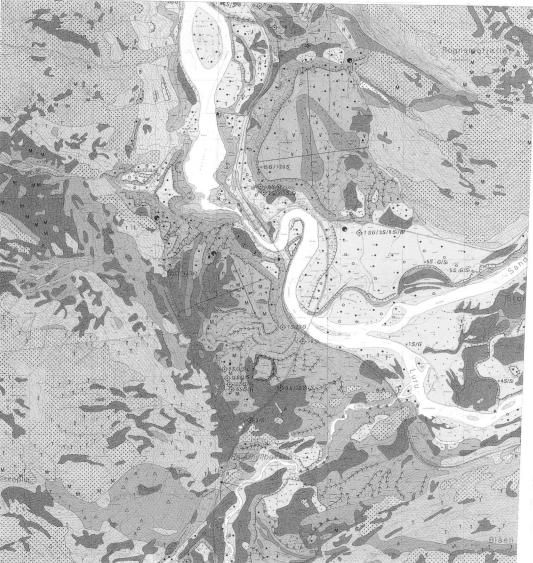

Scale:

0   10 km

Curve interval: 5 m

Legend:

Sediments:

Till, continous sheet of considerable thickness.
Till, discontinous thin sheet.
Marginal -moraine ridge
Glaciofluvial deposits
Marine deposits, continous sheet of considerable thickness.
Marine deposits, discontinous, thin sheet.
Fluvial deposits, postglacial.
Eolian deposits.
Rockfall, talus.
Peat.
Bedrock with a thin and discontinous cover of Antropogen fill.   umus.

Bedrock:
Bare bedrock
Small exposure of bedrock

Patches of sediments on top of dominant deposits of an other origin:

M Till
B Glaciofluvial
H Marine
U Shore
E Fluvial
V Eolian
F Weathered
R Rockfall
J Clay-slide
T Peat

Surface features:
Glacial striae
Glacial river scarp
Meltwater channel
Ice-contact slope
Terrace scarp
Ravine
Clay-slide scarp.
Eolian dune
Abandoned river channel
Gravel pit

Other information:
Drill site
Unit more than 5m thick.

Grain sizes:
Boulders (blocks)
Stones
Gravel
Sand
Silt
Clay

Fig. 3–60. *A general geological map of the Quaternary surface deposits where colors are used to show the genesis of the sediments and symbols show morphological features and sediment textures. The map is a small part of a larger map, and the legend presents a selected number of the symbols and "other informa-tion" which are used on the main map. The legend is translated from Norwegian. (Published with permission from the Norwegian Geological Survey.)*

The map covers the central part of a valley in western Norway. The rather large areas covered with till (light green) suggest that the area was once covered with an ice sheet. The marginal-moraine ridges (dark green), together with a large outwash terrace (orange) of gravel (black dots) on the downvalley side of distinctive ice-contact slopes (red hachured lines), show that the front of a valley glacier was located at this place for a considerable time.

The outwash terraces represent an outwash delta deposited in a shallow fjord. Katabatic (?) winds from the glacier transported and deposited eolian sand (buff color, and V-marks) on the main outwash terrace. The sea entered the lower part of the valley when the glacier retreated, and marine silts and clays (blue) were deposited.

River erosion connected with the subsequent isostatic uplift resulted in several different erosion features, and finally the fluvial sand (yellow) and the peat (brown) were deposited.

This is a brief outline of the general geological history which the map presents. For practical purposes the map gives information about the various properties of the sediments, including grain sizes, and even the thicknesses of some sediment units are indicated. Several of the symbols on the original map are not listed.

used in soil-improvement products and for fuel.

3. *Ground water*. Quaternary sediments are important ground water reservoirs, and many cities, industries, farms, and private households depend upon ground water for water supply. Beds of permeable sediments and zones of fractured bedrock may represent important ground water reservoirs.

4. *Waste disposal and pollution*. Very much of the waste disposal from cities takes place in Quaternary sediments. Since this kind of disposal is a source of pollution, the location of the disposal sites must be selected with care. Ground water is generally transported rather easily through the coarse-grained Quaternary sediments (sand and gravel), and pollution of the ground water may have serious consequences for water supplies and for the water in rivers and lakes. Not only disposed waste, but sewage (from houses, industries and farms) and fertilizers used on farms penetrate Quaternary sediments and are important sources of pollution of the ground water.

However, the sediments provide to some degree a cleaning mechanism for ground water, and several pollutants can be destroyed or trapped as the ground water flows through the sediments. Therefore, it is important to know the source of the pollution, the flow of ground water, and the type and distribution of the sediments to be able to take sufficient precautions against pollution.

Engineering geology

An important aspect of engineering geology is focussed on the Quaternary sediments, which are frequently the ground on which various types of construction are carried out. A thorough knowledge of the sediments, their distribution, thickness and physical properties is needed to do the correct calculations for the construction. The sediments must be surveyed and tested in the field, and samples tested in the laboratory. These kinds of studies, together with calculations for the construction, are generally carried out by engineering geologists. They include traditional geological observations like sediment distribution, grain size and grain-size distribution, grain petrography, water content and content of organic material, for example. In addition several geotechnical tests are made, such as tests of porosity, permeability, shear strength, and sensitivity, depending upon the need.

Land and soil conservation

Considering the economic importance of the Quaternary deposits, and the fact that Quaternary features represent some of the most striking elements of the landscape in many countries, a planned use of, and in some cases a protection of, the deposits and landscape features must have a top priority in both city and landscape planning. Numerous examples could be mentioned of misuse of the ground and sediments as a result of poor planning. Valuable sand and gravel sources have been destroyed, ground water reservoirs ruined, rivers and lakes polluted, and unique landscape features damaged, for example. In most cases an adequate Quaternary mapping is the foundation upon which the planning must be based. Many communities could have avoided both large environmental problems and economic expenses if they had been more careful while doing their basic planning.

Chapter 4
EXTENDED GLOSSARY

This Glossary presents a selection of terms, mainly those that are directly related to the subject of this book, but also some other geological terms that can be useful for a student. Some terms are explained more thoroughly than others, especially the ones dealing most directly with the Quaternary. Several terms which have a rather obvious meaning and many special terms from other fields of geology are not listed. They can be found in dictionaries, and we recommend that students buy one of the inexpensive, paper-back pocket dictionaries listed on p. 195.

A

ablation: the combined effect of all processes by which snow and ice are lost from the glacier. Most ablation takes place at the surface of the land-based glaciers, by means of melting and sublimation; however, it usually takes place mainly at the terminus by calving if the glacier grades into an ice shelf.

ablation zone: the zone on the glacier surface with net ablation, where all snow that accumulates during the winter season melts during the summer season.

abrasion: the wearing away of a rock by rubbing and grinding on its surface.

absolute age: the exact age measured in years (*see* geochronology and relative age).

absorption: the process by which a surface or matter retains radiant energy instead of reflecting or transmitting it.

abyssal: *see* continental shelf.

accelerator mass spectrometer (AMS): an instrument used to measure the mass of different atoms in a sample. The method is much used to find the mass of radioactive isotopes, such as the amount of ^{14}C in organic samples, and in that way date the samples (*see* p. 184).

accessory mineral: a mineral which occurs in very small amounts in a rock or a sediment, such as heavy minerals in some sedimentary rocks.

accretion: enlargement of land which is gained from the sea by the accumulation of sediments or the growth of coral reefs etc. The term is used in other connections also.

Acheulian: a Paleolithic culture (*see* p. 96).

acidic rocks: igneous rocks with more than about 66% SiO_2.

active glacier: a glacier where at least some of the ice is "flowing".

active layer: the upper layer of the ground in permafrost regions, which freezes in the winter and thaws in the summer. Some scientists do not restrict the term to permafrost regions (*see* p. 135).

adhesion: the process by which molecules or particles stick to a surface, caused by forces between the molecules.

adiabatic warming: the warming which takes place in air masses which drop in altitude (flow down a slope). For dry air this warming is in the order of 1°C per 100 m altitudinal drop.

adsorption: the linking of particles mainly by adhesion, such as condensation of gas on the surface of solids, or concentration of dissolved salts on the surface of particles.

aeolian deposit: *see* eolian deposit.

aerobic organisms: organisms which need free oxygen to live; anaerobic organisms do not need oxygen.

aerosols: dust of various kinds in the atmosphere, which lowers the part of the solar radiation which reaches the earth.

agglomerate: a mixture of large and finer-grained rock pieces formed by explosive volcanic action and subsequently petrified.

agglutinated foraminifera: foraminife with shells formed by "foreign" particles which are cemented together.

aggradation: the building up of a surface by deposition.

A horizon: *see* soil.

air gun: energy source used to provide shock waves used in marine seismic surveys (*see* shallow seismic recording).

albedo: expresses the percentage of incoming radiation that is reflected from the earth's surface. The reflection from a snow cover can be as high as 70–90% and from a forest about 10–20%. The corresponding albedo values are 0.7–0.9 and 0.1–0.2, respectively.

alkali-felspar: potassium and sodium felspars, such as orthoclase and albite.

alkaline soils: soils with a pH of 7 or more.

Alleröd Interstadial: *see* Fig. 2–68.

allochthonous rocks and sediments: those which have been moved a "considerable" distance from their original place of deposition.

alluvial fan: a fan-shaped deposit of water-transported material.

alluvial plain: a plain of water-transported material, generally of young age.

alluvium: a fluvial deposit. The term was originally used for a young deposit of Holocene age (*see* diluvium).

Alpine glacier: a glacier in an Alpine mountain terrain.

altimeter: a "barometer" used to measure altitudes.

altithermal: a term which has been used for the warmest Holocene phase. A synonymous term is hypsithermal.

Amersfoort: *see* Fig. 2–17.

AMS dating: *see* radiometric dating.

anaerobic zone: a zone in the sea or a lake where oxygen is lacking (*see* aerobic organisms).

angle of repose: the maximum angle at which the surface of a sediment remains stable.

angular clast: *see* roundness.

angular unconformity: *see* sediment structures.

annual moraines: small moraine ridges oriented transversely to the ice-flow direction, supposedly deposited at the ice front as end moraines with an age difference of one year between successive moraines. However, according to another theory the annual moraines can be crevasse fillings. Annual moraines generally occur in "swarms" of several moraines (*see* washboard moraines).

Antarctic Convergence: the southern equivalent of the Arctic Convergence.

antecedent stream: a stream which has a course independent of present main geologic structures, which were formed by folding, tilting or uplift after the stream course was laid.

anticline: *see* tectonic structures.

antidune: a current ripple of sand which was formed by upstream transport on the stream bed.

antropogenic deposit: a man-made deposit, generally a fill, such as a road fill.

apex of a fan: the top point of a fan.

aquifer: a water-bearing reservoir rock.

aragonite: an unstable $CaCO_3$ mineral which will usually be transformed to the more stable calcite mineral.

archean: ancient.

Arctic Convergence: a zone where the cold Arctic water meets the warmer Subarctic water. The zone corresponds roughly with the oceanic Polar Front.

Arctic species: plant or animal species which live in the cold Arctic zone.

Arctic zone: generally used for land areas to the north of the 10°C July isotherm (*see* p. 30).

arenaceous rocks: rocks which have been derived from sand, or contain sand.

argillaceous rock: a rock composed mainly of clay minerals.

argillite: a mudstone.

arid climate: a climate characterized by dryness.

arkose: a coarse-grained sandstone rich in felspar.

artefact: man-made object of prehistoric age.

artesian water: ground water which is under hydrostatic pressure and is able to rise above the bed containing it.

assemblage zone: a biostratigraphic zone defined by a group of associated fossils.

asthenosphere: the plastic-to-fluid lower part of the mantle, which lies below the lithosphere (*see* earth zones).

Atlantic period: *see* Fig. 2–68.

Atterberg scale: *see* grain size.

auger: *see* soil samplers.

autochthonous: rocks and sediments that have not been moved from the place where they were formed, although they may have been folded and faulted.

avalanche: a large mass of snow/ice or other material which moves rapidly down a mountain slope.

axis: *see* tectonic structures, fabric analysis.

B

backshore: the shore zone above the reach of ordinary waves, above the foreshore (*see* Fig. 4–14).

band: a lamina or bed different in color from adjacent layers.

bar: a term used in many connections. In sedimentology, it means an elongated deposit of sand, shingle or mud in a river channel or in the near-shore part of the sea or a lake. In dynamics, structural geology and glaciology, the term is used for a unit of pressure (air pressure) or stress (shear stress).

barchan dune: *see* dune.

barrier ice: the same as ice shelf.

barrier island: an offshore island, generally long and narrow, often covered with beach ridges and dunes.

barrier reef: a coral reef separated from the coast by a lagoon. The waves generally break on the reef.

basalt: *see* rock.

basal till: *see* till.

base level: the lowest level at which erosion can take place. Generally used in connection with stream erosion where the sea level was considered to be the base level. Today we know that much erosion takes place below sea level (*see* turbidity current). Lake levels and thresholds of very resistant rocks frequently form temporary and local base

levels, which direct erosion and sedimentation in the immediate upstream part.

basement: the flat surface of old igneous and/or metamorphic rocks on which younger, sedimentary rocks rest unconformably.

basic rock: an igneous rock rich in metallic bases, with 45–52% silica.

batholith: a large intrusive body without a visible floor of older rocks (*see* laccolith).

bathyal: *see* continental shelf.

bathymetric chart: a map of the sea floor which shows the water depths by means of contour lines (isobaths).

bauxite: a residual composed of aluminum hydroxide minerals. Bauxite is formed under tropical to subtropical conditions, and it is an important aluminum ore (*see* clay minerals).

b-axis: *see* fabric analysis.

beach ridge: a ridge in the shore zone generally formed by wave action. During storms the ridges can form several meters above sea level, and they are often called storm ridges or storm beaches. Small ridges which resemble "normal" beach ridges can be pushed up by lateral expansion of the ice on frozen lakes.

beach scarp: a steep slope along a beach formed by wave erosion.

bed: *see* sediment structures.

bedded deposit: *see* sediment structures.

bedload: *see* load.

benthonic (benthic) organisms: organisms which live on the sea floor.

Bergschrund: a wedge-shaped opening or crevasse on a glacier, at the head-wall, generally on a cirque glacier (*see* Fig. 3–3).

B horizon: *see* soil.

bicarbonate: a salt containing the anion HCO_3^-.

bi-modale: *see* mode.

biofacies: generally used for an assemblage of animals or plants which lived in a well-defined environment.

biostratigraphic zone (or bio-zone): *see* stratigraphy.

biostratigraphy: a stratigraphy based on fossils and the changes in animal or plant life.

biotope: an area of well-defined ecology.

bioturbation: *see* sediment structures.

bitumen: any natural hydrocarbon, ranging from asphalt to petroleum.

bivalvia: *see* mollusc.

Blancan: the oldest unit (age) of the "Pleistocene" in North America. According to the latest definition of the Pleistocene, the Blancan is of Pliocene age.

block: a clast larger than 256 mm in diameter (*see* grain size).

blockfield: *see* frost features: Felsenmeer.

blow-out: trough-shaped hollow formed by wind erosion. The term is used in other connections also.

bluff: the face of a steep cliff.

bog: a swamp or wetland, mostly covered with peat.

Bölling Interstadial: *see* Fig. 2–68.

boomer: a marine seismic-energy source used in recording the sediment stratigraphy below the sea floor (*see* shallow seismic recording).

Boreal: northern.

Boreal species: plant and animal species which live in the Boreal zone.

Boreal zone: a geographic zone characterized by a special climate, fauna and flora, and which lies between the Lusitanean and the Arctic zones (*see* p. 30).

bottomset beds: *see* delta.

boudinage structure: *see* sediment structures.

boulder: *see* grain size.

boulder clay: a clayey till.

boulder pavement: a flat-lying layer of boulders in a till, generally boulders where the top surfaces are flattened and striated by glacial erosion.

boulder train (boulder trail): glacially transported boulders (erratics) which lie in a zone, usually fan-shaped (or linear), on the lee-side of a well-defined outcrop of a characteristic rock type from which the erratics are derived.

Bouma sequence

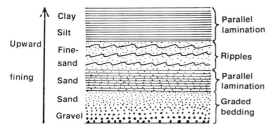

Fig. 4–1. A Bouma sequence is a typical stratification within many density-current deposits. The grain sizes range from gravel to clay in the presented example, but they may range from gravel to silt or sand, of from sand or silt to clay.

Bouma sequence: *see* turbidity current, sediment structures, and Fig. 4–1.

B.P.: before present. Most radiocarbon ages are presented in years B.P., where the year 1950 is defined as present.

Brachiopoda: a marine invertebrate with two unequal shells, called valves.

brackish water: slightly salty water whose salt content is intermediate between fresh and ordinary marine salt water.

braided stream: a stream which flows in a net of changing channels, the cause being that more material is brought into the stream than it is able to transport. This is typical for glacial streams on outwash plains in front of the glaciers (*see* Fig. 3–36).

breccia: a sedimentary rock of angular clasts, of which many are larger than sand size. The angular clasts are usually rock fragments formed by crushing in a fault zone.

Bronze Age: a cultural phase between 4000 and 2500 years old.

Brörup Interstadial: *see* Fig. 2–17.

brown soil: *see* soil.

Bryozoa: small, mostly marine, calcareous, colonial animals.

budget year: in glaciology used about an accumulation season plus the following ablation season.

burrow: *see* sedimentary structures.

C

calcareous: containing $CaCo_3$.

calcite: the crystalline form of the stable mineral calcium carbonate (*see* aragonite).

caldera: the large basin within a circular wall of a volcano.

Caledonian orogeny: *see* Fig. 1–31.

calving: the breaking off and floating away of pieces of ice from an ice front or ice shelf which rests in water.

Cambrian: *see* Fig. 1–31.

canyon: a steep-walled gorge cut by a river.

capillarity: the force which causes water to be pulled up in thin, hair-like "tubes" or interstices in plants, fine-grained sediments etc.

capture: *see* piracy.

carbon–14 (^{14}C): *see* radiocarbon dating, and p. 147.

carbonate compensation depth (CCD): the depth in the ocean below which all carbonate shells are dissolved. The depth is generally close to 4000 m, but it varies somewhat with the temperature of the water.

Carboniferous: *see* p. 35.

catastrophism: a theory which suggests that violent, more or less worldwide catastrophes have been more important than the normal, present-day processes in forming much of the rock and causing regional geologic changes, particularly changes in fauna and flora (*see* uniformitarianism).

cavitation: the corrosive/erosive effect caused by the collapsing of high-pressure bubbles in water (Bernoulli effect).

Cenozoic: *see* Fig. 1–31.

chalk: *see* rock.

chattermarks: *see* glacial-erosion features.

chernozem soil: a soil rich in humus and carbonates which has a thick, black, A-1 horizon (*see* soils). The chernozem is formed under temperate to cool semiarid conditions (*see* Fig. 3–47).

chert: a fine-grained, dense and hard cryptocrystalline siliceous rock (*see* flint).

chlorite: *see* clay minerals.

C horizon: *see* soil.

chronostratigraphy: *see* stratigraphy, geochronology.

chronozone: *see* stratigraphy.

cirque: *see* glacial-erosion features.

cirque glacier: a relatively small glacier which occupies a cirque or part of a cirque. With increasing length of the glacier tongue, the glacier gradually passes into a valley glacier.

clast: a rock fragment. The term is much used in sedimentology as a general term for rock fragments, both rounded and angular rock fragments of various sizes.

clastic: consisting of rock fragments.

clastic dyke (dike): *see* sediment structures.

clast-supported sediment: a sediment where the coarse grains are in contact with and support the structure, in contrast to a matrix-supported sediment (*see* p. 155).

clay: a sediment dominated by clay-size particles, which can be both clay minerals and rock fragments like quartz fragments. Sediments with more than 20% clay-size particles are generally called clays. When wet, the clays are plastic, and when the water content exceeds the liquid limit, the clays become quick and behave like liquids (*see* p. 129).

clay minerals: a group of hydrous-layer aluminum-silicates, which are a major component of clays. These silicate layers consist of two different types of two-dimensional sheets:

1. *the tetrahedral sheet* with linked silicon-oxygen tetrahedra, where aluminum may replace some of the silicon atoms; and

2. *the octahedral sheet* layers with aluminum coordinated with oxygen and hydroxygen ions, where Mg^{2+}, Fe^{2+} and other ions may replace aluminum.

The two kinds of sheets are linked together by oxygen atoms. The assemblage formed by linking one tetrahedral sheet with one octahedral sheet is called a 1:1 layer, while one octahedral sheet sandwiched between two tetrahedral sheets is called a 2:1 layer. The stacking arrangement of the layers, and which of the interlayer cations or octahedral sheets hold the layers together, determine the clay-mineral type.

The four most common clay minerals are:

1. Kaolinite, which has a 1:1 layer structure;

2. Montmorillonite, which has a 2:1 layer structure with exchangeable interlayer cations, and belongs to the smectite group;

3. Illite, which has a 2:1 layer structure with fixed interlayer cations; and

4. Chlorite, which consists of a 2:1 layer with an interlayer octahedral sheet.

Clay minerals develop by alteration of other silicate minerals such as micas, felspars, and amphiboles (and other ferro-magnesium minerals), particularly in the weathering mantle and soil developed on rocks or sediments. Important factors in this process are the mineral composition of the bedrock or clasts, and the climate and drainage, which determine the degree of leaching. *Illite*-formation is common in temperate climates with limited leaching. *Chlorite*-formation represents an intermediate stage of leaching (in temperate climates), preferably in acid soils. *Montmorillonite*-formation also represents an intermediate stage of leaching in temperate climates, preferably in well-drained soils with a neutral pH, and in arid zones. The *kaolinite*-forma-

tion represents a more advanced stage of leaching, and by further leaching and removal of silica, *gibbsite* and other aluminum hydroxides which compose the aluminium ore *bauxite* are formed. In humid tropical climates the red iron- and aluminium-rich lateritic soils are formed through intensive weathering. XRD and DTA analyses are used to identify the clay minerals; see p. 165.

clay size: *see* grain size.

cleavage: the ability of a crystalline rock to split along smooth, more or less parallel planes.

CLIMAP: "Climate, Long Range Investigation, Mapping and Prediction." An international project financed by the U.S. National Science Foundation. The project was in operation during the 1970s and early 1980s, and focussed on marine geology with meteorology, glacial geology, and glaciology.

climatic optimum: *see* hypsithermal period.

climatic zone or climate zone: the globe is divided into several regional climate zones (*see* Fig. 1–28).

climato-stratigraphy: *see* stratigraphy.

clinometer: an instrument used to measure the inclination of beds, planes, axes of clasts etc. (*see* inclinometer).

closed system: a system where no matter enters or leaves during the time under consideration, generally used in reference to chemical systems.

coal: a rock of carbonaceous material derived from former plant/forest vegetation.

cobble: a clast between 64 and 256 mm in size (*see* grain size).

coccoliths: very small calcareous plates which are formed on the surface of flagellate organisms. They are the most common calcareous nannofossils.

cohesion: the force which makes particles stick together in a matter, such as a soil.

coleoptera: beetles, an order of insects. There are numerous different beetle species, and studies of their distribution show that many of them live in very restricted climatic environments. Fossil beetles frequently occur in Quaternary sediments, and beetle stratigraphy has become an important tool in paleoclimate reconstructions (*see* Fig. 2–52).

collapse features and structures: used in glacial geology about features and structures formed by the collapse of

sediments which rested on, or were supported by, ice which subsequently melted. Kettle holes are collapse features (*see* sedimentary structures).

colluvium: a mixture of weathered, or partly weathered, clastic material, mainly transported downslope by gravitational forces.

conchoidal fracture surface: a smoothly curved, glassy surface characteristic of fractured quartz.

concordant: *see* sediment structures.

concretion: a nodular-shaped enclosure in a rock or sediment, usually formed by adding elements from a solution on a nucleus. The concretion is generally harder than the host rock or sediment.

confluence: where two rivers, or two glaciers etc., meet and join.

conformable: *see* sediment structures.

congelifluction: downslope movement of sediments in the active zone on top of the permafrost.

congelifraction: fracturing of the rock by frost action.

congeliturbation: frost action of various kinds in the active zone, including frost heaving and solifluction.

conglomerate: a rock composed of mainly rounded to subrounded clasts of which many are larger than sand size.

consequent stream: a stream which flows in the same direction as the underlying bedrock beds are dipping.

consolidation: the process whereby sediments of any kind become harder and more coherent.

constructional surface: the surface of a deposit which has preserved its original depositional form.

continental crust: *see* earth zones.

continental deposit: a sediment which was deposited within a land area, such as river deposits and lacustrine deposits.

continental drift: the drift of the earth's continents caused by movements in the deeper viscous or fluid parts of the earth.

continental margin: the margin of a continent/continental crust.

continental shelf: the flat part of the sea floor next to a land area, which is no more than 200–400 m deep and ends at the break where the continental slope starts (*see* Fig. 4–2).

continental slope: *see* Fig. 4–2.

continental terrace: *see* Fig. 4–2.

continuous permafrost: *see* frost features.

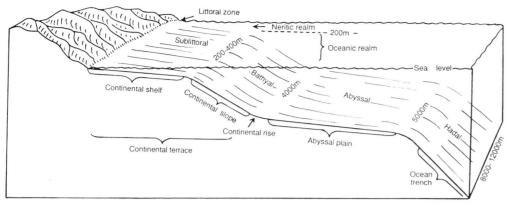

Fig. 4–2. Morphological terms and marine/oceanic terms used in connection with the continental shelves and the ocean floors down to the deepest ocean trenches.

convolute bedding: *see* sediment structures.

core: a term used in various connections (*see* corer, earth zones).

corer: instrument used to collect cores of sediments and rocks, generally cylinder-shaped. A box corer collects square cores.

The following are corers used to collect samples on the sea floor:

1. A gravity corer uses its own weight to penetrate the beds, and it collects cores from the soft sediments, generally no more than 10 m below the sea floor. The original gravity corer is the Kullenberg corer.

2. A vibration corer uses both weight and mechanical vibration to force the sample cylinder down, and it can penetrate much deeper than the gravity corer.

3. A piston corer has a piston which is pulled up inside the cylinder as it penetrates the sediments. The piston is released through a trigger mechanism. Gravity and vibration corers are frequently equipped with pistons.

4. A diamond-bit drill uses a rotating wreath with diamond bits to penetrate the sediments and rocks.

The following are corers used to collect cores from lake or bog sediments (manpower is used to force the corer down into the sediments in most cases):

1. The Livingston corer is a piston corer where the sediment is forced into the cylinder at the same speed as the cylinder is pushed down and the piston is pulled up into the cylinder.

2. The Hiller corer is a chamber corer where the sediment core is collected by opening the side of a generally 0.5–1 m long cylinder (chamber) which has a projecting "knife" along the side. The knife cuts out a sample when the corer is rotated, and the sample is rotated into the chamber.

3. The Russian corer has a half-cylinder-shaped chamber which is closed by a blade that opens when the corer is rotated, and the sediment is rotated into the chamber.

The following are corers used to collect cores below the ground surface on land:

1. Diamond drills are used to collect cores in rocks and sediments.

2. Piston-type corers are used to core soft sediments. However, sediments on land are usually hard to penetrate, and therefore, various motorized devices are used to force the corer-cylinder down.

Many different kinds of samplers, mostly drills, are used to collect samples, generally not cores, from below the ground surface (*see* soil samplers).

coriolis force: a force resulting from the rotation of the earth. The force diverts moving particles, ocean currents etc. to the right in the northern hemisphere and to the left in the southern hemisphere.

correlation: observation regarding corresponding units or beds in two or more stratigraphical sections.

corrie: a less-used term for a cirque.

corrosion: the removing or "eating" away of rock material by chemical action.

crag-and-tail: a streamlined hill formed by a glacier, consisting of a bedrock knob (the crag) with a tail of material, generally till, on the lee-side.

cratons: the large, relatively immobile parts of the earth's crust, also called shields.

creep: very slow downslope movement of material on a slope. The term is used also for a slow deformation of rocks, metals etc. which are subjected to stress over long periods.

crescentic fracture: *see* glacial-erosion features.

crescentic gauge: *see* glacial-erosion features.

Cretaceous: *see* Fig. 1–31.

Cromerian: *see* Fig. 4–7.

crop out: to be exposed at the surface (*see* outcrop).

cross-bedding: *see* sediment structures.

cross-lamination: *see* sediment structures.

cross-stratification: *see* sediment structures.

crust: *see* earth zones.

cryology: the study of snow and ice.

cryomer: a cool or cold phase (*see* thermomer).

cryoturbation: the same as congeliturbation.

crystal: a mineral grain or body with a surface of regularly arranged planes caused by the internal regular arrangement of the atoms.

cuesta: a long bedrock hill in strike direction, steep on one side where the bedrock beds outcrop and gentle on the other, which is a gentle dip-slope.

current bedding (lamination): *see* sediment structures.

current ripple: *see* sediment structures.

D

Daniglacial: the oldest part of the Late Weichselian, when Denmark was deglaciated (*see* Gotiglacial and Finiglacial). The term is not much used today.

Darcy's law: *see* geotechnical parameters.

dating methods: *see* p. 147 and radiometric dating.

Davies' erosion cycle: according to Davies the erosion of a land area is cyclic, each cycle starting with an uplifted high mountain region (the youthful stage), passing through an adolescent and a mature stage, and ending with a low plain near sea level, called a peneplain (the old-age stage).

dead ice: a term generally used about the ice in a glacier which is dynamically dead and so does not move; but it has also been used about ice in a glacier which is climatically dead and has no accumulation area.

dead-ice topography: a hummocky topography with hummocks of till and/or glaciofluvial deposits, frequently with kettle-hole depressions and eskers. The dead-ice topography is characteristic of areas where debris-laden, stagnant parts of a glacier melted (*see* Fig. 2–62).

debris: 1. an accumulation of clasts generally formed by disintegration of rocks; 2. any rock material within or on the surface of a glacier.

debris flow: a "rapid" flow of water-saturated rock debris.

deciduous forest: a forest of trees which shed leaves and are not evergreen.

deckenschotter: *see* schotter.

declination: *see* magnetic declination.

deflation: wind erosion.

degradation: the lowering of a land surface by erosion, mainly by running water.

delta: a unit of river-transported material deposited at the mouth of a river which enters a lake or the ocean. There are several different kinds of deltas. They all have flat delta plains, frequently triangular (fan) shaped, and steeper delta-front slopes. In a cross section the delta consists of flat-lying topset beds over steeper foreset beds which rest on units of bottomset beds that are usually thin and fine-grained. The front slopes of some of the world's largest deltas, where the rivers transport only fine-grained clasts, are very gentle. The typical glaciofluvial deltas usually consist of coarse-grained sediments, and their front slopes are steep with foreset beds generally dipping 10–30°. They are of the Gilbert type (*see* Fig. 4–3, and tidal deltas).

delta plain: *see* Fig. 4–3.

deluge: a term used about a suggested worldwide big flood, generally considered to be the biblical flood.

dendrochronology: a chronology based on counting annual tree-rings and matching rings from different trees/tree-logs, using the obtained tree-ring time scale to date both geological and historical events.

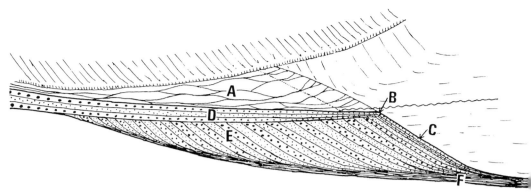

Fig. 4–3. A normal glaciofluvial outwash delta of the Gilbert type.
A: Outwash plain. B: Front break. C: Front slope. D: Topset beds. E: Foreset bed. F: Bottomset beds.

Denekamp: *see* Fig. 2–17.

density current: *see* turbidity current.

denudation: the wearing away of material from the earth's surface by all geological processes and thus the reduction of the altitude of the surface.

desert pavement: a surface layer of coarser clasts formed in a desert as a lag when finer particles are blown away from the surface.

desert polish: smooth, generally "shining" surfaces on clasts and rocks formed by polishing/blasting the surface with sand grains in transport by wind.

desert varnish: a blackish or brownish crust or stain of manganese and/or iron which frequently covers the polished clasts at the surface in some deserts.

Devensian Glaciation: the last glaciation on the British Isles (*see* Fig. 4–7).

Devonian: *see* Fig. 1–31.

diabase: *see* rock.

diachronous unit (zone): a stratigraphic unit which is older in one area than in another; for instance, a time-transgressive unit. The term is used in other connections also, such as a diachronous shoreline, a diachronous end moraine etc.

diagenesis: post-depositional physical and chemical changes in a sediment.

diamict: visually unsorted deposit with a wide range of particle sizes (*see* sorting). Tills are usually diamicts.

diamictite: a petrified diamict.

diapir: *see* tectonic structures.

diatom: a microscopic, single-celled planktonic or benthonic alga with silica cell walls, much used in biostratigraphy and in the determination of sea-level changes (*see* p.144).

diatom ooze: a siliceous mud found on deep-sea floors, consisting partly or mainly of diatoms.

differential thermal analysis: *see* DTA.

diffraction: the change in direction of a wave (ray) when it crosses from one medium into another.

diffusion: a kind of slow mixing of one substance with another.

diluvium: the term has mainly been used for Pleistocene glacial and glaciofluvial deposits. The term is not much used today.

dinoflagellate: a microscopic, unicellar planktonic organism, considered to be a plant.

dinosaurs: two orders of Mesozoic reptiles which comprise the well-known large and spectacular forms.

diorite: *see* rock.

dip: the angle between a surface or bed and the horizontal (*see* tectonic structures).

disconformity: *see* sediment structures.

discontinuous permafrost: *see* frost features.

discordance and discordant: *see* sediment structures.

dissolved load: the part of the load carried by a river which is transported in solution (*see* load).

distal: the opposite of proximal (*see* proximal part or side).

distributary: a branch of a river which flows away from a main branch.

diversity: complexity. In biology a high/low diversity means a high/low number of species.

doline: a circular depression in a karst area.

dolomite: *see* rock.

Donau glacials: pre-Günz glacials in the Alps.

drag fold: *see* tectonic structures.

drainage divide: the boundary between adjacent drainage basins.

Drenthe Stage: *see* Warthe Stage.

drift deposit: any accumulation of glacial and glaciofluvial deposits.

drift sheet: drift deposits in a sheet.

drumlin: a streamlined ridge formed at the base of a glacier and by the glacier. The ridge lies parallel with the flow direction of the glacier, and it consists of subglacial till, sometimes with a core of bedrock or a core of older sediments. The typical drumlin is cigar-shaped with a head and a tail, the tail pointing in the glacier-flow direction (*see* Fig. 3–27).

Dryas: plants of the genus *Dryas* (*see* p. 93).

DTA: abbreviation for Differential Thermal Analysis, which can be used to determine the minerals of clay-size particles. DTA may be used in combination with XRD to determine some clay minerals (*see* XRD).

dune: in geology generally used about an eolian sand dune, which is a hill or bank of sand piled up by the wind. Dunes have different shapes, and they have different orientations with regard to the prevailing wind direction. They have a gentle, wind-facing slope where sand is transported and the surface is eroded, and a steep, lee-slope where the sand is deposited. Some of the best-known dune types are: (1) the barchan dune, which has a curved, lunar shape and the convexity facing the wind; (2) the transverse dune, which is a ridge oriented transversely to the wind direction; and (3) the longitudinal dune, which is a ridge parallel with the wind direction. Coastal dune-belts consist of various hummocky dunes in narrow belts along many sandy beaches. Today the active dune formation occurs mainly in sandy deserts and in zones next to open sandy beaches or sandy river plains which are exposed to periodical, strong, generally dry winds.

dyke: a bed-shaped igneous or clastic intrusion which is discordant with the host beds (*see* sill, clastic dyke, and tectonic structures).

dystrophic lake: a shallow lake rich in organic matter.

E

earth flow: a downslope movement of surface material, somewhat faster than a creep.

earth hummocks: *see* frost features.

earth pillar: a pinnacle, generally of clayey material, capped by a stone and formed by water erosion.

earth's crust: *see* earth zones.

earth's orbit: the elliptical path of the earth around the sun.

earth zones: *see* Fig. 4–4. The different earth zones are determined by means of observed changes in the behavior of seismic and earthquake waves when they pass from one zone into another.

astenosphere: the lower, viscous part of the mantle (*see* Fig. 4–4).

core: see Fig. 4–4.

crust: the upper, rigid part of the earth, above the Mohorovicic discontinuity. The crust below the oceans (oceanic crust) is much thinner than the crust below the continents (continental crust).

Gutenberg discontinuity: see Fig. 4–4.

lithosphere: the solid part of the earth, i.e. the crust and the upper part of the mantle.

mantle: the layer of the earth between the crust and the core.

Mohorovicic discontinuity (the Moho): the seismic discontinuity between the crust and the mantle, about 35 km below the surface of the continents and about 10 km below the ocean floor.

ebb tide: *see* tide.

Echinoderma: a marine invertebrate with carbonate skeleton.

echogram: *see* echo sounder.

echo sounder: an instrument which is used to determine the water depth on the basis of the travel time for sound waves between the instrument and the sea floor. On a moving boat the depths are

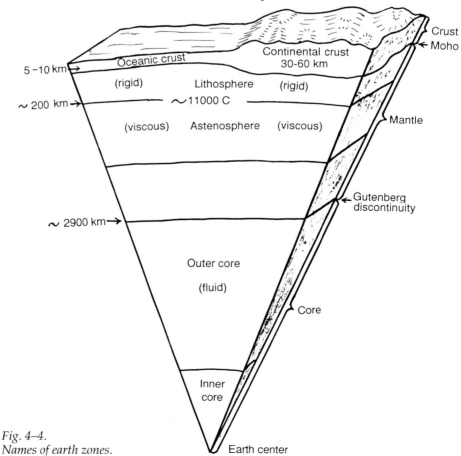

Fig. 4–4.
Names of earth zones.

recorded continuously, and the results are transformed to a graph called an echogram (*see* penetration echo sounder).

ecology: the interrelationship between living organisms and the environment.

ecosystem: an ecological system observed in nature.

ecotone: the limit of a regional bio-zone.

Eemian Interglacial: *see* Fig. 2–1, and p. 44.

effective grain size: the size of a spherical particle which drops in a fluid with the same speed as an observed particle, both particles being of the same specific weight (*see* Stokes' law).

electrical well logging: logging carried out by lowering electrodes in a drill hole and recording the different rock beds' resistance to the induced electrical current.

electron-spin resonance dating (ESR dating): *see* p. 148.

Elsterian Glaciation: *see* Fig. 2–1.

eluviation: movement of substance in a soil by natural soil processes, either in solution or as particles. Soil horizons which have lost substance (A2) are eluviated, and horizons which have received substance (B) are illuviated (*see* soil).

emergence of a coast: the rise-up above the sea level of the near-shore part of a coast, where former shallow sea-covered areas become dry land, because of isostatic or tectonic uplift, or because of a general eustatic lowering of the sea level.

endemic taxon: a plant or animal species which has remained in the area where it originated, without spreading to other areas.

end moraine: *see* moraine.

endogene geology: the part of the geological science which deals with processes that take place within the earth and the results of the processes, including some results shown on the earth surface, such as volcanism and mountain-building.

englacial: within the glacier.

englacial drift: debris which lies within the glacier.

entrenched meander: a stream which lies in a rather deeply incised meander valley.

environment: the sum of all external conditions which influence the life of an animal or a plant.

Eocambrium: the latest part of Precambri-

an, the period when the beds in the Sparagmite Formation in Scandinavia were deposited (*see* Fig. 1–31).

Eocene: *see* Fig. 1–31.

eolian deposit: a deposit which was transported and deposited by the wind (*see* dune loess).

epicenter: the area/point on the earth's surface vertically above the seismic focus where an earthquake originates.

epoch: *see* p. 190.

equilibrium line: the line which separates a glacier's accumulation area from the ablation area. The annual accumulation equals the annual ablation along this line. (*See* snow line.)

era: *see* Fig. 1–31.

erosion: removal of material from any part of the earth's surface.

erosion cycle: *see* Davies' erosion cycle.

erratic: this term is now used about a glacially (including iceberg-) transported rock, generally larger than gravel size. The term is in particular applied to large, isolated erratic boulders on bare rock surfaces. Originally the term was used when the erratic was of a different rock type from the rock surface on which it rests.

esker: a ridge of glaciofluvial material deposited by a glacial river in contact with the ice, either in a subglacial tunnel, which is most usual, or in an open subaerial channel in the ice, or at a retreating ice front which lies in the sea or a lake. The last-mentioned esker type has frequently been called "De Geer esker" or "punkt esker" (punkt = dot), and it may consist of a chain of mounds, small "delta" embryos. When built up to sea or lake level, the mounds are flat-topped, and they are in fact small deltas. If the ice front recedes at an even speed, without marked halts, the De Geer esker can be a smooth esker ridge (*see* p. 131). In Scandinavia the term *ås* or *aas* (plural: *åsar*) has been much used for an esker. In Sweden the term *randås* has been used for a marinely or lacustrinely deposited end moraine consisting mainly of glaciofluvial material (*see* moraine).

esker complex: a complex of esker ridges, frequently combined with kettle holes and kames.

estuary: a drowned, distal part of a former river valley in which the tidewater flows.

eustasy: worldwide changes in world sea level caused by variations in the ocean water volume or size of ocean basins. Most scientists also include changes in sea level caused by changes in the shape of the globe, the geoidal changes, in the term; these include elastic deformation of the earth caused by shifts in load from the ice sheets on the continents to the water in the ocean basins. However, the most dominant factor for the Quaternary eustatic changes is the changing amount of water stored in the large ice sheets.

eutrophic water: a water body which has an excess of plant nutrients derived from human activities.

evaporite: a sediment formed as a result of evaporation, such as salt and gypsum.

exfoliation: the peeling-off of thin rock sheets along planes parallel with the rock surface.

exogene geology: the part of geological science which deals with the processes taking place on the earth's surface and with the results of the processes, except volcanism and tectonism.

extrusion flow: used in glaciology about a glacial flow which has a flow-maximum somewhere below the glacier surface, supposedly caused by squeezing of more viscous layers at some depth within the glacier. Normally this is not possible, and the flow usually decreases with depth below the glacier surface (*see* Fig. 3–3).

F

fabric analysis: analysis of the orientation of clasts in a deposit. In particular, fabric analyses of the coarse clasts in tills, till-fabric analyses, are commonly done (*see* Fig. 4–5).

Usually the direction and plunge of the longest axis (the a-axis) of well-elongated clasts are measured, and clasts larger than 1–2 cm are generally used. However, even smaller clasts can be measured, in particular clasts in oriented samples which are analyzed in the laboratory. In carrying out the analysis the plunge of the intermediate axis (b-axis), or the shortest axis, and the "quality" of the clast (the ratio between the length of the different axes) are frequently noted. If possible, only the most

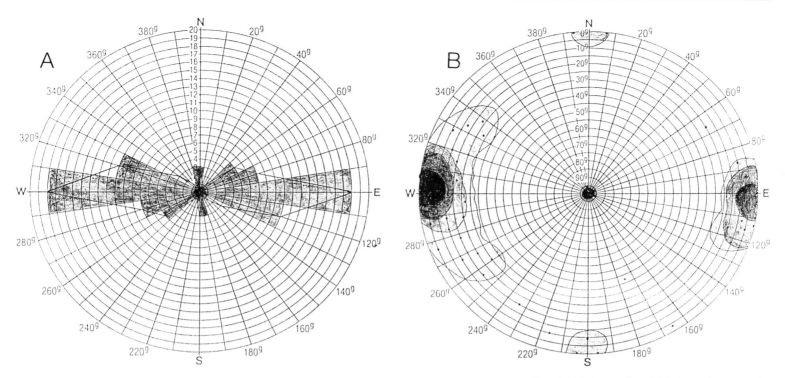

elongated clasts are used. As many as 100 clasts can be analyzed, but generally the analysis is stopped at an earlier stage if the "significance level" is reached (*see* Fig. 4–6).

In the example presented below, 50 clasts were analyzed before the significance level was reached. Figure 4–5A shows a rose diagram, and Fig. 4–5B a Schmidt net plot of the analysis. They indicate that the till was transported in a direction about 100° East. In Fig. 4–6B the Schmidt net plot is contoured.

rose diagram: Figure 4–5A shows how a rose diagram can be constructed. The eastern half of the compass circle is divided into ten 20° sectors, and circles are drawn representing the number (or percentage) of clasts. The figure shows an analysis of 50 clasts which have the following a-axes directions: 18 clasts within the 90° to 110° sector, 10 clasts within the 110° to 130° sector, 7 clasts within the 70° to 90° sector, 5 within 50° to 70° sector, 3 within 130° to 150°, 2 within 150° to 170°, 3 within 190° to 210° and 2 within 10° to 30°. The mirror image of the eastern half is generally added on the western half of the circle to complete the rose.

Schmidt net: This is an azimutal projection on a horizontal plane of the lower hemisphere of a sphere, which is used to plot directions and plunges of axes; for instance, of a-axes of clasts. Each dot on the projection plane records both direction and plunge of the axis.

facies: the term is used in lithostratigraphy about the general appearance or character of unit sediments or sedimentary rocks, in particular the character relating to the environment in which they were deposited; for instance, marine facies, eolian facies etc. (*see* also mineral facies).

fanglomerate: a petrified fan deposit.

fault: *see* tectonic structures.

Number of clasts within the primary mode:

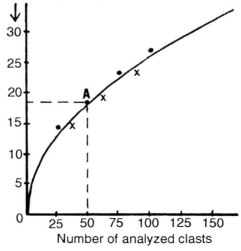

Number of analyzed clasts

Fig. 4–5. The results of fabric analyses can be presented as rose diagrams (A) or contoured Schmidt nets (B). A represents the same analysis as B, but B is superior to A since it shows both the direction and the dip of the a-axes. B shows a marked majority of a-axes dipping in westerly direction, which indicates that the transport direction was towards the east. A does not indicate which of the two directions was the transport direction. A rose diagram can be presented in two ways, as indicated in A. Note that the circles are divided in 400 so-called new degrees. The numbers on the circles in A represent numbers of clasts; for instance, 18 clasts were oriented in east-west direction (within the 90° to 110° sector).

Fig. 4–6. Fabric analysis of till.
The graph shows how many clasts must be analyzed to obtain a significant orientation strength.
The graph represents minimum orientation strength based on the 95% level of probability (modified from Harris, 1969).
The number of analyzed clasts is plotted against the number of clasts within the primary mode of orientation. If the plot lies on or above the graph (dots), then the significant orientation strength is reached and there is a 95% statistical probability that the mode represents the correct transport direction of the sediment.
A: The plot of the analyzed sample in Fig. 4–5.
Dots: Samples with significant orientation strength.
X: Sample without significant orientation strength.

faunal province: an area characterized by an assemblage of animals which are widespread within the area.

Felsenmeer (or blockfield): *see* frost features.

felsic minerals: light-colored minerals like quartz, felspars, felspathoids, muscovite.

felsite: a fine-grained igneous rock composed mainly of quartz and felspar.

felspar: *see* mineral.

Finiglacial: the time when most of Finland was deglaciated (*see* Gotiglacial).

firn: very coarse-grained "snow" which has a density between 0.4 and 0.82, generally more than 1 year old.

firn line: the lower limit of firn on a glacier surface during that part of the melting season when this limit lies at its highest. It corresponds roughly with (or lies slightly below) the equilibrium line.

fission-track method: *see* p. 149.

fjärd: *see* glacial-erosion features.

fjord (fiord): *see* glacial-erosion features.

Flandrian: a term used for the same time period as Holocene.

flexure: *see* tectonic structures.

flint: a variety of chert which occurs as nodules, generally in chalk. Flint is very tough, has a conchoidal fracture with sharp edges, and was therefore very much used by Stone-Age man to make implements.

flocculation: the forming of particle complexes by joining suspended particles in a fluid to "larger" lumps. The process is generally promoted by electrical forces.

floodplain: the flat plain on either side of a stream which is periodically flooded.

flow limit: *see* geotechnical parameters.

flow till: *see* till.

flume: this term is generally used for a man-made "channel" used in a laboratory for experiments with flowing water and transport of sediments.

fluted surface: *see* glacial-erosion features.

flute mark (cast): *see* sediment structures.

fluvioglacial: a less used term for glaciofluvial.

flysch: synorogenic fine-grained sandstones, shales, marls etc. of mid-cretatious to lower-Tertiary age which lie on both flanks of the Alps. The flysch consists very much of turbidites deposited in the marine environment in geosynclines. The term has been used for similar deposits in other areas also.

föhn wind: a wind which has passed over a mountain and flows down the leeward slope, where it gradually becomes warm (adiabatic warming) and dry.

fold: *see* tectonic structures.

foliation: a layering caused by the arrangement of mineral grains in a rock.

foraminifera (abbr. forams): small, unicellular, mostly marine animals with mainly calcium-carbonate shells, generally microscopic. There are both benthic and planktonic species. Studies of forams are much used in the biostratigraphy of marine sediments.

forebulge: *see* isostasy.

foreset beds: *see* delta.

formation: besides the obvious general use of the term, it has a special meaning in stratigraphy (*see* p. 189). The term "formation" has also been used in naming stratigraphic sections.

frequency distribution: numerical distribution of objects (for instance, minerals) in different classes.

frontal moraine: the same as terminal moraine and end moraine (*see* moraine).

frost features: all features formed by frost action, such as frost mounds, polygons and wedges (*see* p. 135).

active layer: the upper layer of the ground in a permafrost region; the layer which thaws during the summer season. However, the term has been used in a wider sense to include the freeze-thaw layer even in areas outside the permafrost regions.

Felsenmeer (or blockfield): a jumble of angular, frost-shattered rock fragments of all sizes, broken from the underlying rock surface (*see* Fig. 3–41).

frost heaving: the lifting of the sediments in the active layer caused by freezing of pore-water and formation of ice-lenses when water freezes at the freezing plane. Since marine clay has a high pore volume and silt/fine sand has a high capillarity, these kinds of sediments are usually most influenced by frost heaving. Coarse-grained sediments are usually well drained, and with no water and no frost heaving. However, even coarse-grained sediments can be poorly drained, water-saturated, and subjected to heaving, particularly in permafrost regions.

ground ice: layers, lenses or more irregular bodies of pure ice within the ground in the permafrost zones.

Ground-ice layers can be several tens of meters thick (*see* p. 137).

upfreezing: also called "frost shooting" of stones or other objects within the annual freeze-thaw layer (the active layer). The stones etc. are moved by the frost towards the surface. In fact they are moved directly towards the freeze plane, which is usually vertically upwards. Two main theories have been proposed for this process: a "pull theory", which suggests that the stone is pulled up with the freeze layer, and a "push theory", in which the stone is pushed up by the ice which forms below the stone. The latter theory is based on the fact that the stone transmits cold better than the sediment. Observations indicate that both processes take place.

ice-wedge cast: the wedge-shaped sediment filling in a wedge which was formerly filled with ice. Ice wedges and ice-wedge casts can be wide at the top and very deep. However, wedges which are 0.5–2 m wide and 1–3 m deep are most common. Ice-wedge casts are formed in permafrost regions. Flat-lying beds are usually bent up at the contact with the ice wedge (*see* Fig. 3–42), but when the ice melts they are partly bent down, or faulted down into the cast. (*See* sand-wedge casts, p. 169.)

non-sorted circles, nets, polygons, steps, stripes etc.: see sorted circles, etc. on p. 169.

pals (plural: palsar): a mound of organic deposits with a core of ice formed in bogs in areas with discontinuous permafrost. A pals is generally 1–5 m high (*see* Fig. 3–46).

patterned ground: a ground with patterns of polygons, circles, stripes, mounds etc. caused by frost (*see* Figs. 3–44, 3–45).

permafrost: permanently frozen ground. Permafrost areas can be divided into areas with continuous permafrost and areas with discontinuous permafrost, which represent transition zones to areas without permafrost. The southern limits of present-day zones of continuous and discontinuous permafrost in the northern hemisphere correspond roughly with mean annual isotherms of minus 6°C to minus 8°C and minus 1°C, respectively (*see* Fig. 3–40).

pingo: a large mound consisting of beds

of inorganic sediments, generally silt and sand overlying a core (lenses) of ice, formed in areas with discontinuous or thin continuous permafrost. There are two types of pingos, the closed system and the open system. The former has been called the "Mackenzie Delta" type, and it is formed on flat plains or frequently in former shallow-lake depressions within areas with continuous thin permafrost. The latter is formed on valley floors or gentle valley slopes. Pingos can be more than 30 m high, and a depression with a lake is usually formed on top of a pingo during the collapse phase. Fossil pingos are generally circular, low ridges with a shallow depression in the center (*see* Fig. 3–46).

sorted circles, nets and polygons: frost features where coarse clasts (gravel to boulders) are concentrated along the frost fissures and form patterns which are visible on the ground surface. The coarse clasts have been frozen up ("sorted out") out of the original sediment. According to the most accepted theory they are frozen up to the surface from which they creep laterally down into the cracks, and some stones may have been moved directly to the upper part of the crack. There are in fact several theories presented to explain the detailed mechanism behind these kinds of frost features. Non-sorted features have the same patterns as the sorted, but they have no concentration of coarse clasts along the fractures. They form in polar desert regions or on sediments which contain no coarse clasts (*see* Fig. 3–45).

sand-wedge casts: wedge-shaped casts formed by freeze processes in permafrost regions and filled with sand. They are formed in dry polar desert regions, and the sand is generally eolian. Flat-lying beds adjacent to the cast are usually bent up at (near) the contact with the cast (*see* p. 138).

G

gabbro: *see* rock.

gamma-ray well logging: logging of boreholes by observing the radioactivity of the rocks through which the holes pass.

Gastropoda: a class of the phylum *Mollusca*, mainly known as snails.

gelifluction: the kind of solifluction which takes place both in areas with permafrost and in areas with only seasonal frozen ground.

genus (plural: genera): a group of species with fundamental characteristics in common.

geochronology: the field of geological science which is focussed on the exact age of the geological deposits and events, as measured by radiometric methods, varves, tree-rings etc. Since there are various problems involved in determining the "exact" age by these methods, it is now common to identify the method by which the age has been found, such as varve years, radiocarbon years.

geocryology: the term is mainly used about the study of frozen-ground processes and features, but it has also been used in a wider sense for the study of both frozen ground and glaciers.

geoid: the shape of the globe.

geomagnetism: the earth's magnetic field and related phenomena (*see also* paleomagnetic pole).

geomorphology: the part of the geological and geographical sciences which deals with land forms and their formation.

geosyncline: a narrow, subsiding belt of the earth in which thick units of sediments accumulate.

geotechnical: relating to the geological aspects of engineering.

geotechnical parameters and tests of sediments are used to record physical/mechanical properties of the sediments. Some important parameters/tests are briefly described in the following:

Porosity is expressed as the ratio of pore volume over sediment volume.

Void ratio is expressed as the ratio of pore volume over volume of solids.

Relative density of a granular sediment is based on comparison between the *in situ* void ratio and the void ratios for the densest and the loosest possible packing of the same sediment.

Permeability (hydrolic conductivity) is the ability of a sediment to transmit a fluid (generally water) through the pores and their interconnections. A permeameter is frequently used to record the permeability. It may consist of a tube, 10 cm^2 in diameter, filled with the sediment. The water is "forced" through the tube at a given pressure gradient, and the water flow is measured. Notice that a very porous sediment is not necessarily very permeable. For instance, marine clay is very porous but almost impermeable, while a much less porous sand can be very permeable.

Darcy's law: a law which describes the flow (when laminar) of fluids through a porous sediment. It states that the rate of flow is proportional to the driving force.

Plastic limit is measured by adding water to a clayey sample until it becomes barely so plastic that it can be shaped (rolled) to a 3.2 mm thick string. The water content (percentage by weight) of the sample at this stage represents the plastic limit.

Liquid limit of a clayey sample is an expression for how much water the sediment can hold before it starts to "flow". A Casagrande instrument or a conus instrument can be used to determine the liquid limit. Water is added until the sediment receives the necessary plasticity for a given degree of flow. The water content of the sample at this stage represents the liquid limit.

Plasticity index for a clay is an expression for the plasticity, i.e. the zone in which the clay is plastic. The difference between the liquid limit and the plastic limit is the plasticity index.

Deformation modules present the ability of a sediment to resist deformation caused by imposed stresses. Laboratory oedometer tests (one-dimensional situation) or triaxial tests may be used to determine deformation moduli for different stress conditions.

Shear strength: the internal strength or resistance of a sediment (or a rock) against being sheared. The shear strength can be measured in the laboratory on an undisturbed sample by means of various methods, including the laboratory triaxial test. In clayey sediments it can be measured in the field by means of a vane test in which a rod with wings is rotated in the sediment and the resistance is measured. The shear strength is related to the effective stresses in the sediment, i.e. the total stresses minus the pore-water pressure.

British Isles	N. Germany Holland	Poland	European Russia	Siberia
Devensian	Weichselian	Vistulian	Valdaian	Sarta Zyrjanka Karginsk Int. Ermakavo
Ipswichian Intergl.	Eemian	Eemian	Mikulinoian	Kazantsevo
Wolstonian	Warthe Saalian Trene Int. Drenthe	Wartanian Saalian Pilica Int. Odranian pre Odranian	Dnjeperian	Tazovian Bachta Shirta Int. Samarovo
Hoxnian Intergl.	Holsteinian	Mazovian Ferdynandovian	Likhvinian	Tobol
Anglian	Elsterian	Sanian Nidanian	Okaian	Shaitan
Cromerian	Cromerian	Przasnyszian Podlasian	Berezinaian	Talagaika
Beestonian	Menapian	Narevian	?	?
Pastonian	See Fig. 2-1B			

Fig. 4–7. Names of glacials and interglacials used in different regions; see also Fig. 2–1. The use of names for the oldest glacials and interglacials is not consistent in some countries and regions; The presented names for Poland are based on information from D. Krzyszkowski (personal communication).

geothermal gradient: the change in temperature with depth, generally expressed as degrees per unit depth. A rise in the order of 2–3° per 100 m in the upper earth crust has commonly been recorded.

Gilbert delta: a delta with coarse-grained sediments which has a steep delta front-slope (*see* glaciofluvial delta).

glacial: a term used generally for features and deposits related to glaciers and glaciations, but also used for a stratigraphic time period such as between two interglacials or between the last interglacial and the Holocene, for instance the Weichselian Glacial. Figure 2–1 on p. 38 presents the names used for glacials in the Alps, North America, and northern Europe. Corresponding names for glacials in some other regions are presented in Fig. 4–7.

glacial drift: material transported by, or in transport by, glaciers or glacial rivers. Also, glaciomarine deposits have frequently been included in the term.

glacial-erosion features:

chattermarks: crescentic scars on pebbles or glacially eroded bedrock surfaces.

cirque: a large, steep-walled incision in the side of a mountain, usually amphitheater-like, formed by head-wall erosion (quarrying) of a cirque or valley glacier (*see* Figs. 3–16, 3–17).

crescentic fracture: also called "parabolic fracture", is a parabolic-shaped fracture on the rock surface formed below an active glacier. The convexity is facing in upglacier direction, and crescentic fractures frequently occur in swarms of several fractures (*see* Fig. 3–19).

crescentic gauge: also called lunar fracture ("sigdbrudd" in Scandinavia), is an upglacier, concave incision on the rock surface formed below the base of a glacier which "chipped out" a piece of rock (*see* Fig. 3–19).

fjärd: a glacially sculptured open sea arm with gentle, generally not parallel sides, and a bottom with shallow glacial troughs.

fjord (fiord): a glacially sculptured narrow sea arm with steep, relatively high and semi-parallel sides and a trough-shaped bottom (*see* Fig. 1–5, and *Sognefjord* on p. 171).

fluted surface: in glacial geology, used when describing a surface with narrow parallel ridges, generally less than 1–2 m high, but in some cases higher. The ridges were formed below a glacier, and they are parallel with the glacier-flow direction. They can be either rock ridges or till ridges (*see* p. 124).

glacial grove, glacial striae, glacial polish: straight-lined marks of various depths on the rock surface formed when the glacier sole with different size rock fragments slides on the rock surface. The marks (lines) are parallel with the ice-flow direction.

glacial striation analysis: analysis of various glacial-striation directions to determine the relative age of the different ice-flow directions. For instance, the oldest striation direction is frequently found on the lee-slope of glacially sculptured rock knobs. If two sets of crossing striation directions occur on a rock surface, then the youngest set generally consists of thin (polished) striae on the "highs" between the deeper-eroded older striae.

hanging valley: a tributary valley which has a floor at a considerably higher level than the trunk valley at their junction. Hanging tributary valleys are characteristic for glaciated valley systems (*see* p. 115). However, hanging valleys can form in non-glaciated regions also, under special geological conditions, but they do not have U-shaped cross profiles.

lateral channel: also called marginal channel ("spylrenne" in Swedish). A meltwater channel formed along the contact between the lateral glacier margin and the valley side.

lee-side: see stoss-sides, below.

nivation: erosion connected with a perennial snowbank, caused by frost action and mass movement.

p-form or plastic form: shallow smooth grooves, winding smooth channels on a rock surface, formed underneath a glacier, supposedly by abrasion of sediment-loaded water under high pressure and speed (*see* p. 118).

plough marks: shallow, generally 1–10 m deep, erosion marks on the sea floor formed by ploughing of icebergs in the sea-floor sediments. Plough marks frequently occur as swarms of semi-parallel "furrows" on the shallow marine shelves next to glaciated, or formerly glaciated, coastal regions.

plucking: in glacial geology the term means quarrying of rock blocks or fragments from the lee-side of a rock at the base of a glacier (*see* p. 117).

roche moutonnée: also called whaleback or sheep-back rock: a glacially sculptured bedrock knob with a gentle stoss-side and a steep lee-side, formed by, respectively, glacial grinding and glacial plucking. The size can vary from a few centimeters to several meters (*see* Fig. 3–18).

Sognefjord: A fjord in Norway, more than 1300 m deep. The deepest in the world outside presently glaciated regions. (The Nordvestfjord in Greenland is about 1460 m deep.) (*See* Fig. 3–10.)

stoss-sides: in glacial geology used about slopes that have been facing the flow of an overriding glacier; the opposite of lee-side.

trough: a term used in many different connections, also in glacial geology for a closed depression with a smooth "U-shaped" cross profile. Glacial troughs are characteristic forms in glacially sculptured landscapes. They can be from a few centimeters to several hundred meters deep and up to many kilometers long and wide, and they can be of regular or very irregular shape.

The sides of some troughs are gentle and some are steep (*see* fjord and *fjärd,* p. 170).

U-shaped valley: a valley with a smooth "U-shaped" or more or less "U-shaped" cross profile formed by glaciers. This is the normal profile of a glacially sculptured valley.

whale-back: see roche moutonnée.

glacial erratics: *see* erratic.

glacial geology: the branch of geology which is focussed on the results of glacial activity, such as erosion and deposition, rather than the physics of the glaciers, which is the main topic in glaciology.

glacial grove, glacial striae, glacial polish: *see* glacial-erosion features.

glacial theory: *see* p. 12.

glaciation limit, also called glaciation threshold: this is the lowest altitude at which glaciers can form. The altitude is determined by the climatic condition of the area. Mountains with peaks above the glaciation limit generally carry glaciers, and mountains with peaks below this limit do not carry glaciers. The altitude of the glaciation limit is calculated as the mean between the peak-altitude of the lowest-lying mountain with a local glacier and the peak-altitude of the highest-lying mountain with no glacier. A reasonable number of mountains with tops near (above and below) the altitude of the glaciation limit is needed to do exact calculations. In northern Norway this limit lies about 100 m to 300 m above the equilibrium line.

glacier: a body of snow, firn, and ice which flows or has flowed. According to Ahlmann's classification, there are temperate, subpolar, and high-polar glaciers. The ice in a temperate glacier is at the pressure melting point, except at the surface during the winter. No melting generally takes place on the surface of a high-polar glacier (*see* p. 107).

glacier flow: *see* p. 107.

glacier sole: the debris-loaded lower part of a glacier (*see* p. 110).

glacio- or glacial-: used in combination with fluvial, lacustrine and marine to indicate the relationship to glaciers and glaciations.

glaciofluvial delta: a delta deposited in the sea or a lake by glacial rivers, which generally carried much coarse material.

The delta has rather steep foreset beds, frequently dipping between 10° and 30°, and it is of the Gilbert type (*see* Fig. 3–37).

glaciolacustrine deposits: glacial-lake deposit.

glaciology: the science dealing with glaciers and their regime, dynamics, and character.

glaciomarine deposits: *see* p. 127.

gneiss: *see* rock.

Gondwanaland: the supercontinent in which all the continents in the southern hemisphere, and India, were joined until about 180 million years ago (*see* Fig. 1–22).

Gotiglacial: the time when southern Sweden was deglaciated, before the deposition of the Middle Swedish (Younger Dryas) moraines; the time between Daniglacial and Finiglacial. The term is not much used.

graben: *see* tectonic structures.

graded: (1) a graded slope has a profile adjusted for transportation of the surface material. (2) A graded stream channel has a gradual descending longitudinal profile adjusted to transport the existing bed load, and it is graded to a base level, generally a temporary base level such as a lake level or a threshold of hard rock. (3) A graded bed shows a gradation in grain size generally from coarse below to finer above.

gradient: the steepness of a slope, a line etc.

grain size: the two most used classification scales for grain sizes of clasts are the Atterberg scale and the Wentworth scale (*see* Fig. 4–8 and p. 151). Figure A presents the Wentworth and Phi-scales in the normal way, with increasing phi-values and decreasing grain sizes towards the right. Figure B presents the traditional Atterberg scale with increasing grain sizes towards the right. For comparison the Wentworth and phi-values are added. The Atterberg scale is based on the metric system and the number 2 for the grain-size limits. The Wentworth scale is adapted to the Atterberg scale, but with phi-values for the grain-size limits. Therefore, the Wentworth limits are fairly close to the Atterberg limits.

The Phi-scale was developed to simplify numerical grain-size calculations. The phi-value of a grain is the negative

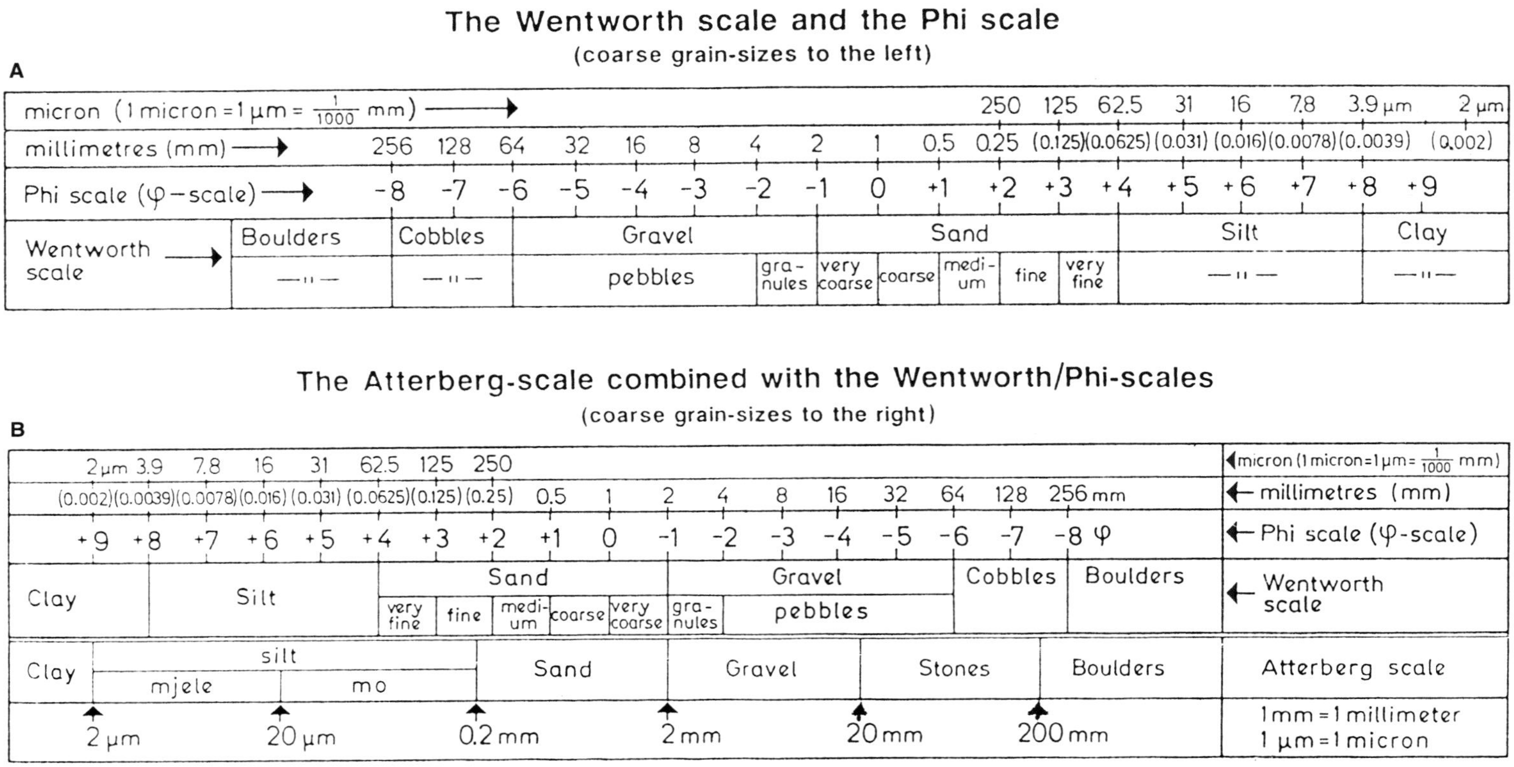

The Wentworth scale and the Phi scale
(coarse grain-sizes to the left)

A

micron (1 micron =1 μm = 1/1000 mm) ──→											250	125	62.5	31	16	7.8	3.9 μm	2 μm
millimetres (mm) ──→	256	128	64	32	16	8	4	2	1	0.5	0.25	(0.125)	(0.0625)	(0.031)	(0.016)	(0.0078)	(0.0039)	(0.002)
Phi scale (φ−scale) ──→	−8	−7	−6	−5	−4	−3	−2	−1	0	+1	+2	+3	+4	+5	+6	+7	+8	+9
Wentworth scale ──→	Boulders	Cobbles	Gravel				Sand							Silt		Clay		
	—"—	—"—	pebbles				gra-nules	very coarse	coarse	medi-um	fine	very fine		—"—		—"—		

The Atterberg-scale combined with the Wentworth/Phi-scales
(coarse grain-sizes to the right)

B

2 μm	3.9	7.8	16	31	62.5	125	250									◄ micron (1 micron=1 μm= 1/1000 mm)
(0.002)	(0.0039)	(0.0078)	(0.016)	(0.031)	(0.0625)	(0.125)	(0.25)	0.5	1	2	4	8	16	32	64	128	256 mm ← millimetres (mm)
+9	+8	+7	+6	+5	+4	+3	+2	+1	0	−1	−2	−3	−4	−5	−6	−7	−8 φ ← Phi scale (φ-scale)
Clay	Silt					Sand				Gravel				Cobbles	Boulders	← Wentworth scale
					very fine	fine	medi-um	coarse	very coarse	gra-nules	pebbles					
Clay	silt			Sand		Gravel		Stones		Boulders	Atterberg scale
	mjele	mo									1 mm =1 millimeter / 1 μm =1 micron
2 μm	20 μm		0.2 mm		2 mm		20 mm		200 mm		

Fig. 4–8. Grain-size scales presented in the "normal" way (A) and in the "Scandinavian" way (B). The Wentworth grain classification is based on the Phi-scale (Udden scale) and the grain sizes increase towards the left. The Atterberg grain classification (B) is the original classification which has increasing grain sizes towards the right. The two ways of presenting the grain sizes are reflected in the ways of presenting grain-size distribution diagrams also.

logarithm (with base 2) of the grain size (in millimeters), which is the negative value of the exponent with base 2 which gives the grain size in millimeters. Examples:

Grain size 1 mm has a phi-value of 0 ($2^0 = 1$).

Grain size 8 mm has a phi-value of −3 ($2^{+3} = 8$).

Grain size 0.5 mm has a phi-value of +1 ($2^{-1} = 0.5$).

A grain-size distribution (gsd) graph shows the weight percentages of the different grain sizes in a sediment sample. The procedures used in collecting and analyzing a sample for grain-size distribution studies will be briefly described:

Sampling: If the sediment contains large clasts (boulders etc.), then only the matrix is sampled and only clasts smaller than about 25 mm are generally used for the analysis. The coarser clasts are separated by sieving.

Sieving: the method most used to separate grain sizes larger than silt (larger than 0.0625 mm) (see p. 151). Sieves with full phi-size or $\frac{1}{4}$ phi-size nets are generally applied. Dry-sieving is used for samples which contain very little or no clay, and wet-sieving is used for clayey samples. The finer-grained fraction, which passes the +4 phi sieve, is collected, dried, weighed, and used for the continued analysis.

Analysis of the silt and clay fractions: Several different methods can be applied to record the silt and clay fractions, such as the hydrometer, pipette, falling drop, sediment weight, laser and sedigraph methods. Most methods are based on Stokes' formula, which expresses the speed in which spherical particles smaller than 0.2 mm drop in a fluid (see p. 129). The pipette and hydrometer methods have been much used, but they are fairly slow. Instruments based on other methods are rather expensive, but they give quicker results, and they may register continuously all grain-size fractions. Some of them are connected with print-ers, which print out continuous cumulative graphs.

Presentation of the grain-size distribution: The different ways in which the grain-size distribution can be presented are shown on page 152. Cumulative graphs represent the most accurate way to present the gsd of a sample.

Cumulative graphs: The weight percentage of any grain fraction of a sample can be read from a cumulative graph. As already mentioned (p. 152), there are two methods which are much used to present cumulative graphs, one with a normal arithmetic percentage scale and the other with a probability percentage scale. A logarithmic grain-size scale is used in both methods. An advantage with the log-probability graphs is the fact that the different grain populations are recorded as more or less straight lines.

Interpreting cumulative graphs: Grain-size distribution can be diagnostic or semi-diagnostic in interpreting the environment in which the sediment was transported and deposited. The following are some examples. Both the arithmetic and the log-probability graphs are presented for each sample to show the difference between the two meth-

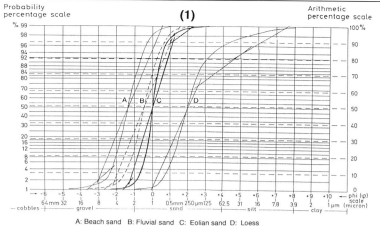

A: Beach sand B: Fluvial sand C: Eolian sand D: Loess

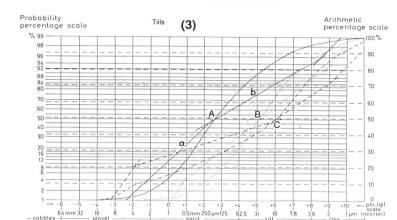

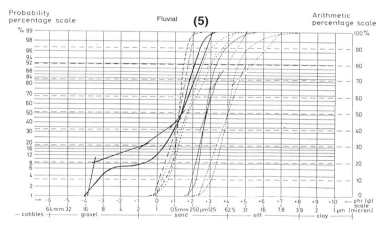

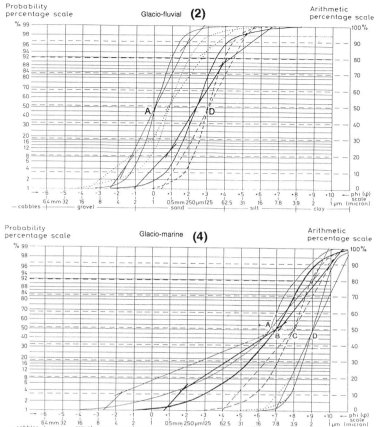

Fig. 4–9. Examples of some typical grain-size distributions in different kinds of sediments.

Graphs based on the probability percentage scale (straight lines) and the arithmetic percentage scale are constructed for each sample. Note that the former graphs consist of broken lines, which usually suggests that each sample consists of several grain populations. The glaciomarine samples (4) represent ice-proximal to ice-distal samples from A to D. Note the change in sorting.

One of the fluvial sand samples (5) has a rather marked gravel population, which represents a "rolling member".

The graphs represent samples from Scandinavia. They are selected, and do not show the complete variability in grain-size distribution of the different kinds of sediments.

ods. Note also that the log-probability graphs are not complete. For practical purposes they start at 1% and end at 99%. This is done because the scales for the first and the last percentages are very long (*see* Fig. 3–58). However, the "full scale" was used for the construction of the presented graphs.

Example 1.
Figure 4–9 (1) presents characteristic graphs for beach sand (A), fluvial (river) sand (B), eolian sand (C), and loess (D). Note that all log-probability graphs are broken lines, suggesting that each sediment consists of several grain populations, as indicated in the figure text. The percentages of the different grain populations can vary depending upon current conditions and the availability of grain sizes when the sediment was deposited. Some sediments may consist of only one population; they are log-normal. Graphs A, B, and C represent sand deposits, which are the most common eolian, river, and beach deposits. However, coarser-grained river and beach deposits are rather common also. The range in median diameter and sorting of the different deposits is indicated in Fig. 4–11.

Example 2.
Figure 4–9 (2) presents a series of rather typical graphs for glaciofluvial deposits. Graph B represents a gravelly coarse

sand which was deposited close to the ice contact. Note that this deposit is considerably less sorted than the sediment represented by graph D, which was deposited at further distance from the ice margin. Most graphs show a small suspended-load population, and some a rolling-grain population.

Example 3.

Figure 4–9 (3) presents some rather typical grain-size distributions for tills. They are all poorly sorted. The log-probability graphs are broken lines which represent different grain populations. Graph A has two dominant modes, one in the sand fraction (a), and the other in the silt fraction (c). They resulted in this case from glacial grinding of rocks with several different kinds of minerals and grains. The different populations in the other graphs may reflect glacial erosion of subglacial sediments of different grain sizes.

Example 4.

The graphs in Fig. 4–9 (4) represent glaciomarine sediments deposited at increasing distances (A to D) from a grounded ice front. The glaciomarine deposits are frequently about as poorly sorted as tills (A and B), but they generally contain much more fines than the tills. However, some glaciomarine sediments can be better sorted (D). The most fine-grained populations (clay-silt) of the presented graphs represent the suspended sediments, as well as sediments suspended in connection with currents. The coarser-grained populations (sand-silt) could have been deposited by bottom currents, by turbidity currents or by drop-out from surface jet-currents or icebergs.

The graphs presented in Examples 1 to 4 are typical graphs, but they do not represent the many possible variations of graphs within the mentioned categories.

Grain-size parameters: The grain-size parameters are used to express numerically different, important characteristics of the grain-size distribution and different features of the cumulative graphs. The four most used parameters are: median diameter, sorting, skewness, and kurtosis. Unfortunately there are several different methods used to

calculate the parameters, and the values obtained can be fairly different. Therefore, the grain-size parameter values must be accompanied by information about the method used. One way of doing this is by using the special nomenclature (special letters) which are connected with the applied method, such as the ones used in the "Folk and Ward" method, shown in the following:

$$\text{Mean diameter: } M_z = \frac{\phi_{16} + \phi_{50} + \phi_{84}}{3}$$

$$\text{Sorting: } \sigma_I = \frac{\phi_{84} - \phi_{16}}{4} + \frac{\phi_{95} - \phi_5}{6.6}$$

$$\text{Skewness: } Sk_I = \frac{\phi_{16} + \phi_{84} - \phi_{50}}{2(\phi_{84} - \phi_{16})} + \frac{\phi_5 + \phi_{95} - 2\phi_{50}}{2(\phi_{95} - \phi_5)}$$

$$\text{Kurtosis: } K_G = \frac{\phi_{95} - \phi_5}{2.44(\phi_{75} - \phi_{25})}$$

Note that phi (ϕ) values are used in all formulas. If the sediment is very fine-grained, with a high percentage of grains finer than 10 phi, then the calculation of ϕ_{95} and even ϕ_{84} can be problematic if the grain-size analysis was stopped at the 10-phi level (which is usual). In that case the ϕ_{84} and ϕ_{95} percentages must be calculated by extrapolation.

The median diameter is usually close to ϕ_{50} (*see* Fig. 4–10). A scale for the sorting is presented in Fig. 4–11. The skewness of a log-normal sample, expressed by one straight line on the log-probability paper, is zero. Samples with fine-grained "tails" (graphs C and D in Fig. 4–9 (1)) have positive skewness, and samples with coarse-grained "tails" (A in Fig. 4–9 (1)) have a negative skewness. The kurtosis values increase with increasing distances between ϕ_{95} and ϕ_5 and decreasing distances between ϕ_{75} and ϕ_{25} (*see* Fig. 4–10).

Plotting of parameter values: Calculated values of the different grain-size parameters can be plotted on triangle diagrams or on two-axial diagrams, and in some cases rather diagnostic patterns are obtained. However, the diagrams obtained must be used with care. The plotting of sorting against median grain-size is much used, and Fig. 4–11 shows a generalized diagram where zones for the different genetic sediments are outlined. The diagram is based on numerous observations and it

was made to give a rough indication of the position of the zones. The diagram shows a considerable overlap between the zones, and other criteria must be used to determine the exact origin of the sediments which plot in overlapping areas.

granite: *see* rock.

granulometry: the measuring of grain sizes.

gravel: *see* grain size.

gravimeter: an instrument used to measure the gravity. Gravity properties of the earth's crust are measured and interpreted by means of gravimetry. For instance, the thicknesses of some glaciers have been measured by this method.

gravity: the force acting upon an object pulling it in a direction towards the center of the earth.

gravity corer: *see* corer.

greenhouse effect: the effect of the atmospheric content of water vapor, ozone, carbon dioxide and some other gases, which allows much of the short-waved radiation from the sun to penetrate to the earth's surface, but absorbs or reflects back to the earth's surface most of the long-waved radiation from the earth.

ground moraine: *see* moraine.

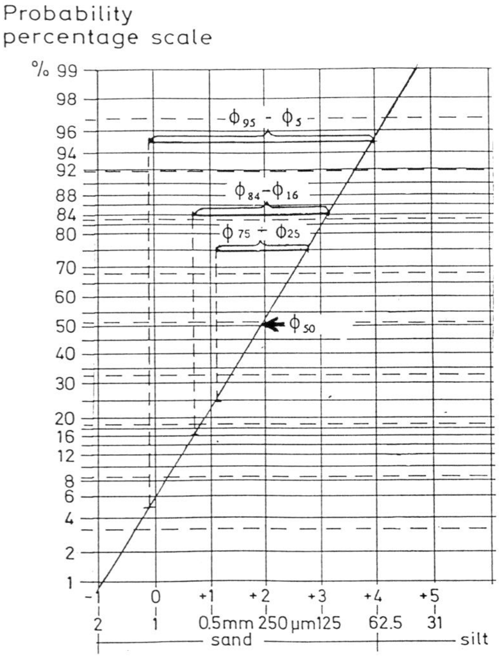

Fig. 4–10. Values used in the presented grain-size parameters. A simple straight-line (single population) distribution graph is used in this example.

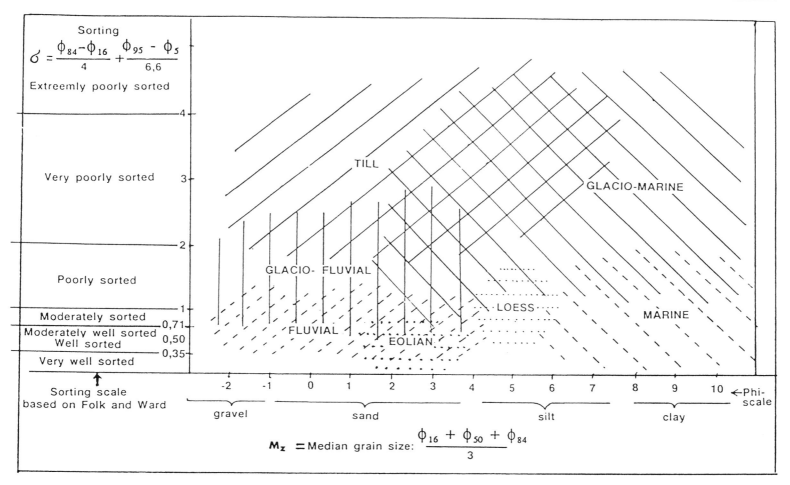

$$\sigma = \frac{\phi_{84}-\phi_{16}}{4} + \frac{\phi_{95}-\phi_5}{6,6}$$

Sorting

Extreemly poorly sorted

Very poorly sorted

Poorly sorted

Moderately sorted

Moderately well sorted
Well sorted

Very well sorted

Sorting scale
based on Folk and Ward

←Phi-scale

gravel   sand   silt   clay

$$M_z = \text{Median grain size: } \frac{\phi_{16}+\phi_{50}+\phi_{84}}{3}$$

Fig. 4–11. Median grain size plotted against sorting.

Plot-areas for genetically different sediments are outlined. The limits for the plot-areas are not very accurate since the diagram is based on relatively few analyses. Note that there is a considerable overlap between the different plot-areas.

ground water: the water in the saturated zone below the ground surface. The ground water fills cracks and pores in both bedrock and sediments. The water supply for many districts and cities depends upon ground water. Some ground water wells can yield much water, particularly wells in coarse sediments. Pollution from dissolved matter can destroy the ground water for much human use.

ground water divide, ground water flow, ground water reservoir, ground water table: terms used in connection with ground water. The divide is a line which separates two reservoirs, and the ground water table (surface) is the upper level for free ground water. This level corresponds with the water level in a bore hole or well in the ground.

guide fossil: *see* index fossil.

gumbo: a sticky, leached and deoxidized clay formed in a soil.

gumbotill: a strongly leached and chemically decomposed sticky, clayey till.

The term is mainly used in North America.

Günz: *see* p. 38.

Gutenberg discontinuity: the boundary between the earth's core and mantle (*see* earth zones and Fig. 4–4).

gypsum: an evaporite mineral of hydrous calcium sulphate.

gyttja: a black mud composed mainly of tiny, partly microscopic plant remains. The gyttja is a common lacustrine sediment.

H

habitat: the environment in which a plant or animal lives or lived.

hadal: *see* continental shelf.

half-life of an isotope: the time it takes for a radioactive isotope to disintegrate one-half of its radioactive atoms.

hanging glacier: a small, thin glacier on a steep mountain slope (*see* Fig. 3–7).

hanging valley: *see* glacial-erosion features.

hardpan: a hard layer formed by precipita-

tion of iron hydroxide or manganese in the B horizon, below the A2 horizon in a podsol profile (*see* soil).

heavy minerals: minerals that have a specific weight greater than that of bromoform, 2.85.

Heinrich layers: layers rich in ice-rafted debris found in some cores from the North Atlantic. The layers record periods with increased discharge of icebergs, probably derived from the ice sheet in eastern Canada. They are poor in foraminifera, and correspond with periods of marked cooling of the surface water. The approximate ages of the observed layers are: 69 000, 52 000, 35 500, 26 500, and 14 500 years B.P.

heliophilous plants: plants which demand much light or full sunlight.

hematite (haematite): an iron mineral (Fe_2O_3), the principal iron ore.

Hengelo: *see* Fig. 2–17.

herbs: seed plants with no woody tissue, where the part above the ground generally dies during "winter season".

Hercynian orogeny: a late Paleozoic orogeny.

hiatus: *see* sediment structures.

high Arctic: the Arctic area where the average air temperature for the warmest month is below 0°C, according to Nordenskjold.

histogram: a graph where the percentages of different elements, for instance the different grain sizes in a sediment, are presented by vertical bars (*see* Fig. 3–58).

Holocene: the latest of the two Quaternary epochs; the last 10 000 years of the earth's history (*see* p. 35).

Holsteinian Interglacial: *see* Fig. 2–1.

Hominidae: the name of a family which has two subfamilies: (1) Australopithecidae (semi-man) and (2) Hominidae or Homo (real man) (*see* p. 96).

Homo: *see* Hominidae. The term homo is used in several connections meaning "the same", the opposite of hetero.

homopycnal: *see* jet flow.

Hooke's law: the deformation of an elastic material is proportional with the applied stress, when used within the elasticity limit.

horizon: *see* sediment structures.

hornfels: a hard fine-grained rock of quartz, felspar, mica etc. formed by contact metamorphism.

horst: *see* tectonic structures.

Hoxnian Interglacial: the second-youngest interglacial in the British Isles (*see* Fig. 4–7).

humic acids: organic acids formed mainly by decay of organic material, in particular plants.

humidity: the content of water vapor in the air. Relative humidity is the ratio between the existing amount of water vapor and the maximum amount that the air can hold at the existing temperature. Cold air can hold much less water vapor than warm air. Therefore, cold air can be very dry, have a low humidity, but the relative humidity can still be very high.

hummock: a mound or knob. Hummocky topography is a much-used descriptive term.

humus: the brown-to-black, more or less decomposed organic matter in the A_1 horizon of soils and swamps (*see* soil).

hyaline: glassy.

hydrogeology: the geology related to the occurrence and flow of ground water, and the use of ground water as water supply.

hydrometer: an instrument by which the weight of suspended particles of different sizes in a fluid (water) can be measured. The hydrometer consists of a closed glass tube, with a scale, that sinks into the fluid. It can be used to determine the weight of the different fractions of clay and silt in a sample which is suspended in a fluid (generally water with a dissolved disperser). This is called the "hydrometer method".

hydrosphere: the water on the earth's surface.

hydrostatic pressure: the pressure at any given level in a water body at rest.

hydrothermal water: water heated by igneous/volcanic activity.

hydroxide: a chemical compound where the ion OH^- is included.

hyperpycnal: *see* jet flow.

hypopycnal: *see* jet flow.

hypsithermal period: a term generally used for the postglacial climatic optimum, the warmest Holocene period. However, the term has been used for the warmest period of any late Cenozoic interglacial also.

I

ice: a solid state of water with a specific weight of about 0.9.

ice age: a glacial period in the earth's history when large glaciers covered areas which are ice free today. Best known are the ice ages during the last 2.5 million years (*see* p. 24).

iceberg: slabs of floating ice broken off the front of grounded glaciers or ice shelves. Some icebergs are several kilometers wide.

ice cap: a more or less dome-shaped glacier which is smaller than an ice sheet.

ice front: the front (terminus) of a glacier.

ice sheet: a glacier which covers a wide area. The ice sheet may have outlet glaciers in valleys and fjords. The term is generally used for very large glaciers.

ice shelf: the floating part of a glacier or many merging glaciers.

ice tongue: the tongue-shaped distal part of a glacier.

ice wedge: a vertical wedge-shaped body of ground ice which extends from the surface into the ground in regions with permafrost (*see* frost features).

ice-wedge cast: *see* frost features.

igneous rocks: rocks formed from molten material, at or below the earth's surface. The contrast is the other big group of rocks, the sedimentary rocks.

Illinoian Glacial: *see* Fig. 2–1.

illite: *see* clay minerals.

imbricated clasts: *see* sediment structures.

inch: one inch = 2.54 cm.

inclination: *see* tectonic structures.

inclinometer: an instrument used to measure the inclination, such as the plunge of axes and the dip of beds. Geologists generally use compasses with built-in inclinometers (*see* clinometer).

index fossil: (1) a fossil characteristic of, and restricted to, an assemblage zone, or (2) a fossil selected to give a name to a bio-zone. The term guide fossil is frequently used also.

infiltration: a term much used for the slow "flow" of a fluid into the pores and cracks of a dry ground (soil) or a ground saturated with ground water. The fluid can be dissolved chemicals, polluted water or plain water.

inland ice: a less-used term for an ice sheet.

INQUA: International Union for Quaternary Research, has more than 50 participating countries. INQUA is promoting Quaternary research in various fields, for instance by organizing international congresses every fourth year, sponsoring commissions etc. (*see* IUGS).

in situ: in place; said about rocks, sediments, fossils etc. which lie in their original place of deposition, and which have not been moved out of place.

insolation: received solar energy (radiation) on the earth's surface.

interbedded: *see* sediment structures.

interglacial: the time between two glacials, when the climate, the regional vegetation zones and the glacial conditions on the earth's surface were approximately as they are today (*see* p. 44).

interstadial: a term generally used in connection with the Quaternary glaciations, meaning a relatively warm phase between colder phases (stadials). An interstadial is usually cooler and shorter than an interglacial. However, an interstadial can be as warm as an interglacial, but in that case it must be of so short a duration that the glaciers of the world were not reduced to interglacial sizes, and the regional vegetation zones did not reach the extent of the interglacial zones.

intrenched (entrenched) stream: a stream which flows in a deep, narrow valley that has been cut as a result of rejuvenation. When such a stream is meandering, it forms entrenched meanders.

invertebrate: animal without a backbone.

inverted: beds that are turned upside down (*see* tectonic structures).

involution: *see* sediment structures.

Ipswichian Interglacial: the last interglacial on the British Isles (*see* p. 170).

isobar: a line or surface through points of equal air pressure.

isobase: a line through points on the earth's surface with equal isostatic uplift; a line through points where raised shorelines (etc.) of the same age lie at the same altitude.

isobath: a contour line (a bathymetric line) through points on the sea floor with equal depths.

isochron: a line through points, generally places on the earth's surface, related in the way that time-transgressive geological events, or other events, happened at the same time in the points on the line (*see* isoline).

isoclinal fold: *see* tectonic structures.

isoline: a line through points which have an equal value of an attribute, for instance a line through places with the same marine limit (ML) (*see* isochron and isoplet).

isoplet: a line or surface through points which have an equal value of an attribute, such as temperature, air pressure, uplift etc.

isostasy: the state of balance created as the lithosphere (the earth's crust and upper mantle) is "floating" on the upper asthenosphere (the viscous part of the mantle) (*see* Fig. 4–4).

isostatic anomaly: the difference between the observed value of gravity and the theoretically correct value.

isostatic compensation: isostatic adjustments which take place when changes in the load on the crust in an area is compensated by vertical movements of the lithosphere.

isotherm: a line connecting points with the same temperature.

isotopes: atoms of the same chemical element which have the same number of protons in their nuclei but a different number of neutrons.

isotropic: having the same properties in all directions.

IUGS: International Union of Geological Sciences. The main geological union of which most countries in the world are members, and with which several "branch unions" are associated, such as INQUA.

J

jet flow: a term used for the flow of high-altitude strong winds, and for the extended flow of rivers into lake/ocean water which has a markedly different density than the river water. Depending upon the density differences, the river jet can be a surface current (hyperpycnal) or bottom current (hypopycnal). If the lake/ocean water is stratified, then an intermediate current may occur. When there is no significant difference in density between the river water and the lake/ocean water, no jet flow develops, and the river water mixes instantly with the lake/ocean water. This flow is called homopycnal; see Fig 4–12.

joint: fracture in rocks or sediments, more or less transverse to the bedding, along which no slip has taken place (*see* tectonic structures).

Jökulhlaup: Icelandic for a sudden flood caused by the drainage of an ice-dammed lake, a lake at the margin or within a glacier. The Icelandic ice-dammed lakes result from melting caused by volcanic activity.

Jura: *see* Fig. 1–31.

K

kame: a large mound or small hill of glaciofluvial sand and/or gravel deposited in a basin on the ice, or a chamber in or below the ice. The kame may contain irregular beds of till and have a thin cover of till or erratics.

kame-and-kettle topography: a complex of kames and kettles, usually a part of a dead-ice topography.

kame terrace: a terrace of mainly glaciofluvial material deposited by a river which flowed along the side of a valley glacier, generally a stagnant glacier (*see* p. 132).

Kansan: *see* p. 38.

kaolin: a rock or a clay composed mainly of clay minerals of the kaolinite group.

kaolinite: *see* clay minerals.

K/Ar age: *see* p. 148 and radiometric dating.

karst topography: a topography characterized by sinkholes, caves and underground tunnels and chambers caused by the solution of limestone, dolomite

JET FLOWS

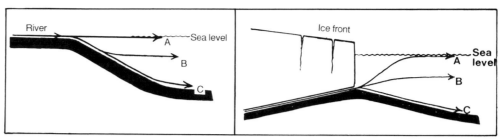

Fig. 4–12. A: Surface jet flow B: Intermediate jet flow. C: Bottom jet flow.
Surface jet flows occur when the river water is lighter than the basin water, which is common when the basin water is marine. Intermediate jet flows may occur when the river water is heavier than the surface water and lighter than the bottom water. Bottom jet flows occur when the river water is heavier than the basin water, which is common when the river contains cold, muddy glacial water and the basin water is fresh or brackish.

or gypsum in underground water bodies and streams.

katabatic wind: a wind that flows downslope from a high-pressure area towards a peripheral lower-pressure area. The cold, dry and heavy air over a high glacial plateau may obtain a very high speed and a very low relative moisture content as it flows downslope. At the same time the air becomes adiabatically heated (*see* p. 68).

kettle: a term used in glacial geology about a depression in drift, most usual in glaciofluvial material, formed by the melting of a body of ice which was totally or partly buried in the drift. The depression is frequently bowl-shaped (*see* kettle hole).

kettle hole: a term generally used for a kettle, also called a kettle depression (*see* Fig. 3–36).

Kullenberg corer: *see* corer.

kurtosis: *see* grain-size parameters.

L

laccolith: a mushroom-shaped (dome-shaped) igneous intrusion.

lacustrine sediment: a sediment deposited in a lake.

lag: a term used about both a time delay and a lag deposit.

lag deposit: a residuum of coarse clasts, the remains of an original deposit which also contained finer clasts that were removed by wind or water action.

lagoon (laguna): a body of shallow water between the shore and a longshore reef, or a body of standing water between the shore and a barrier island.

lahar: a diamicton composed of mainly volcanic ash which was transported downslope as a mud flow.

lamina: *see* sediment structures.

laminar flow: a low velocity flow of a fluid (water) or air; a flow in parallel sheets. The opposite of a turbulent flow.

lamination: *see* sediment structures.

landslide: a "body" of rock or soil which slides relatively fast downslope.

late glacial: a term used about a late phase of deglaciation of an area. In northern Europe some scientists have used this term with a capital L (Lateglacial) to designate a period from about 13 000 years ago to about 10 000 years ago.

lateral channel: *see* glacial-erosion features.

lateral moraine: *see* moraine.

laterite: the material in a latosol.

latosol: a highly leached soil rich in iron, silicium and/or aluminum, generally very red, formed in humid tropical climates (*see* soil).

lava: *see* rock.

leach: to remove soluble minerals etc. from soil, sediment or rock by percolating water.

lee-side: (1) the downglacier side of bedrock knobs and hills. The opposite of stoss-side (*see* glacial-erosion features); (2) the downwind side of any protruding object, such as hills, rocks, dunes, houses.

lenticular: lens-shaped.

levee: a ridge-shaped bank of sediments along either side of a channel, usually a river channel, deposited when the water overflowed the sides of the channel.

lichenometry: age calculation based on the size and growth rate of lichens on rocks.

lignite: a brownish or brown coal in which the alteration of the plant material has proceeded further than in peat and less far than in black coal.

limb: *see* tectonic structures.

limestone: a rock consisting of calcium carbonate.

limnic deposit: the same as lacustrine deposit.

limnology: the science focussing on lakes.

lithification: the process in which a sediment is changed to a rock.

lithofacies: the character of a rock or sediment which describes the sedimentary environment in which it was originally formed.

lithology: the physical/mineralogical character of a rock or sediment, including grain size, sorting and structures.

lithosphere: *see* earth zones.

lithostratigraphy: *see* stratigraphy.

Little Ice Age: *see* p. 96.

littoral zone: shore zone.

load: used in connection with a water stream, the term means the total load transported by the stream, including the dissolved load, the suspended load and the bed load or traction load. The suspended load consists of small clasts distributed throughout the water column. The bed load consists of larger clasts transported in contact with the

stream bed by rolling, sliding, and saltation (bouncing).

load cast: *see* sedimentary structures.

loam: a mixture of fine-grained clasts and organic matter in a soil.

lobate moraines: moraines deposited along the margins of glacier lobes.

lobe: a term used in many connections, for instance for a tongue-shaped part of a glacier (glacier lobe) or a soliflucted soil (solifluction lobe).

lodgement till: *see* till.

loess (löss in German): an eolian silt which usually contains some clay and fine-sand particles, generally non-stratified. Loess is a very common ice age deposit (*see* p. 68).

longitudinal dune: *see* dune.

loss of ignition: *see* p. 155.

Loveland loess: an Illinoian loess in USA.

Lusitanian zone: the zone with Lusitanian flora and fauna which lies to the south of the Boreal zone (*see* Fig. 1–28). The term has been used for a lower Jurassic stratigraphic zone also.

lutite: a sediment or a sedimentary rock dominated by clay-size particles.

lysocline depth: a zone where dissolution of carbonate shells takes place, immediately above the calcite-compensation depth.

M

maar: a depression formed by a volcanic explosion and occupied by a lake.

macro: means large, visible without use of a microscope.

mafic rocks: dark-colored igneous rock composed of non-felsic minerals.

magma: molten rock material.

magnetic declination: the angle between the magnetic and the geographic meridians through the point of observation.

magnetic inclination: the angle between the direction of the earth's magnetic field-lines and a horizontal plane at the point of observation.

magnetic susceptibility: a medium's (sediment etc.) susceptibility to magnetism (*see* stratigraphy).

mammal: a warm-blooded vertebrate animal that brings forth babies and suckles them.

mammoth (*Elephas primigenius*): an extinct long-haired elephant (*see* p. 69).

mantle: *see* earth zones.

marginal channel: *see* glacial-erosion features: lateral channel.

marine limit (ML): also called highest shore level ("högsta kustlinjen, HK" in Sweden), is the highest-lying postglacial marine shore level at any point on a coast that has been glaciated, or the highest-lying shore level occupied during the Quaternary in areas outside the glaciated regions. The ML corresponds with the mean sea level, and storm beaches may lie above the ML. The upper limit on observed wave-washed zones above ML is frequently called the wash-limit (skyllegrense or svallgrense in Scandinavia).

marker horizon, or marker bed: a bed or horizon with a characteristic, easily recognizable sediment, which has a wide lateral distribution. Some ash beds are typical marker beds (*see* p. 20).

marl: a calcareous clay, usually with shell fragments.

marsh: a flat, low and wet ground which may be permanently wet with more or less open water, or periodically dry, and with a variable vegetation.

mass balance: in glaciology used when discussing the balance between accumulation and ablation on a glacier.

mass spectrometer: an instrument used to measure the mass of various atoms in a sample, such as oxygen isotopes, carbon isotopes ($^{12}C - ^{13}C$), uranium isotopes etc. With some instruments the mass of the isotopes can be measured so accurately that the results are used in age determination. (*See* radiometric dating.)

mass wasting: a downslope movement of material caused by gravitational forces.

matrix: in geology used when discussing the finer-grained material which lies between and surrounds the larger grains in a rock or a sediment.

mature soil: a well-developed old soil, supposedly in equilibrium with its environment.

meander: a sinuous bend of a river.

mean sea level: the mean between high and low tide observed during a long period.

mechanical weathering: disintegration (breaking up) of rock surfaces and clasts by mechanical processes such as heating, cooling, freezing, thawing.

medial (median) moraine: a moraine formed between two joining ice streams as a seam of debris along the contact between the ice streams. Frequently the lateral moraines on the joining sides of the two ice streams grade into the median moraine at the junction of the streams.

megaripple: *see* sedimentary structures.

member: a term used in lithostratigraphy for a unit within a formation (*see* p. 189).

Menapian: *see* Fig. 2–2.

Mesozoic: *see* p. 35.

metamorphic facies: *see* mineral facies.

metamorphism: the process by which consolidated rocks change in texture and composition as a result of external forces like heat, pressure and introduction of new chemical substances. Note that weathering and diagenesis are not included in the process.

metasomatism: the change in mineralogy of a rock resulting from chemical interactions caused by migrating fluids.

meteorite: an object of rock or metal which has reached the earth's surface from outer space. Man-made objects are not included in the term.

mica: a mineral group consisting of phylosilicates with sheet-like structures. The best-known mica minerals are the Biotite and the Muscovite.

micro-facies: the character of a sediment or sedimentary rock as seen in a thin section, including lithology, mineralogy, texture, and fossil character.

micro-fauna: a fauna of small animals which must be identified under a microscope.

micro-fossil: a small fossil which must be identified under a microscope.

micron: the length of 0.001 mm, generally written 1 μ or 1 μm.

Mikulino Interglacial: the last interglacial in Russia (*see* p. 170).

Milankovitch: *see* p. 29.

Mindel Glacial: *see* p. 38.

mineral: a naturally formed element found in rocks which has a narrow range in chemical composition and usually a characteristic crystal form. There are numerous types of minerals, such as quartz, different kinds of felspar and mica, pyroxen, hornblende, olivin etc. For further information see textbooks, pp. 195–96 (and *see* clay minerals, rock).

mineral facies and metamorphic facies: rocks that have originated under temperature and pressure conditions so similar that they contain roughly the same minerals and are said to represent the same mineral/metamorphic facies.

Miocene: *see* Fig. 1–31.

misfit stream: a stream which is much too small to have eroded the valley in which it flows.

mixed-layer clay minerals: *see* clay minerals.

mode: in sedimentology the term is used for a dominating grain-size class of a sediment. A bi-modal sediment has two dominating grain-size classes.

Moershoofd: *see* Fig. 2–17.

Mohorovicic discontinuity (the Moho): *see* earth zones.

molasse: a Swiss term for late- and post-orogenic deposits of Tertiary age, found in the foothills of the Alps, formed of short-transported detritus from the uplifted mountains, and consisting of mainly coarse-grained sandstones and conglomerates but also some marls, the former mainly deposited in fresh water and the latter in marine environments. The term has also been used for deposits of similar origin in areas other than the Alps.

mollusc (or mollusk): an invertebrate animal of the *Mollusca* phyllum (*see* taxonomy). In Quaternary biostratigraphy there are two classes of molluscs that are of particular interest, the *Gastropoda* (snails) and the *Bivalvia* (mussels and clams).

monadnock: a residual hill projecting up from a flat denudation surface.

monocline strata: *see* tectonic structures.

monsoon: a dominating wind in the Asiatic-Australian region. In the summer the wind blows landwards from the ocean (wet monsoon), generally in a northeasterly direction. In the winter it blows seawards from the continent (dry monsoon), generally in a southwesterly direction.

montmorillonite: *see* clay minerals.

moraine: the term has been used in different ways, but it is now mainly used as a morphologic-genetic term for a body of glacial material generally dominated by till. There are several kinds of moraines, such as subglacial moraines (also called ground moraines or basal moraines), supraglacial moraines (also called ablation moraines), marginal moraines, of which there are two kinds, lateral moraines and end moraines (also called frontal moraines and terminal moraines) and finally medial moraines (*see* p. 119). Marginal moraines are frequently ridge-shaped. End moraines deposited in the sea are commonly stratified and consist both of till beds and beds with sorted glaciomarine and glaciofluvial-type sediments (*see* p. 126). In Fennoscandia the term "morene" (in German-speaking countries, Moräne) has been used both as a geomorphological term and as a term for the unsorted deposit, the till, which generally composes most of the moraines. This has complicated the nomenclature. It is, for instance, problematic to use the term end moraine for a marinely deposited end moraine which consists of mainly glaciofluvial or glaciomarine-type sediments, and not of much "morene", here meaning till. Therefore, the term "randås" (marginal esker) has been used in Sweden, the term "randbildning" (marginal formation) has been used in Finland, and the terms "randdannelse" (marginal formation) and "brefrontrygg" (ice-front ridge) have been used in Norway. (*See* also till.) In some areas even terrestrial end moraines, in particular push moraines, may consist of mainly pushed-up glaciofluvial deposits.

morphology: the science dealing with the surface features of landscapes, their shape and how they are formed. The term is used also for the surficial shape of sediment grains and fossils.

morphostratigraphy: *see* stratigraphy.

mud: a generally wet, fine-grained material of mainly clay-size particles, often mixed with organic material in a soil.

mud cracks: cracks formed when wet mud dries.

mudflow: a flow of water-saturated mud, frequently containing an admixture of larger clasts.

mylonite: a rock of small, crushed rock fragments formed along thrust planes etc.

N

nannofossil: a very small micro-fossil such as a coccolith.

Nansen bottle: a bottle used to sample water at different depths.

nappe: *see* tectonic structures.

natural selection: a process by which less-adapted individuals tend to be eliminated from a population, leaving no descendants.

Nebraskan Glacial: *see* Fig. 2–1.

Neogene: the latest part of the Cenozoic, after the Oligocene, after about 26 million years ago (*see* Paleogene).

Neoglacial events: events which occurred in postglacial time, in Europe after about 8000 years ago.

Neolithic: the new Stone Age; the part of the Stone Age which in Europe is younger than about 5000 years.

neotectonic: recent and subrecent tectonic.

neritic deposits: shallow-water deposits, generally on the continental shelf, deposited below low-tide level and above about 200 m depth.

net ablation: a term used when the annual ablation on a glacier is larger than the annual accumulation.

nivation: *see* glacial-erosion features.

non-sorted circles, nets, polygons, steps, stripes etc.: *see* frost features.

normal distribution: in sedimentology, used about a grain-size distribution which plots as a Gaussian graph. For instance, the grain-size distribution in a well-sorted sediment with one grain population is usually normal (*see* p. 172).

nunatak: an isolated peak or hill with exposed bedrock which projects up through the surface of a glacier and is surrounded by glacier ice.

O

obsequent: an obsequent stream flows transversely across outcropping bedrock beds that dip in the direction opposite to the stream-flow direction. The opposite of the flow direction of a consequent stream.

obsidian: a glassy rhyolite.

ocean-floor spreading: *see* plate tectonics.

Odderade: *see* Fig. 2–17.

offlap sequence: *see* sediment structures.

offshore bar: a sand ridge built under water in the breaker zone (*see* Fig. 4–14).

Older Dryas: *see* p. 80.

Oldest Dryas: the cold phase and the corresponding vegetation zone, immedi-

ately before the Bölling Zone (*see* p. 93). The lower boundary of the Oldest Dryas is not well defined.

Oligocene: *see* Fig. 1–31.

onlap: *see* sediment structures.

oolite: usually a limestone composed of small, rounded, concentrically layered grains.

ooze: a deep-sea sediment, either calcareous and deposited between about 2000 m and 3900 m below sea level, or siliceous and deposited below about 3900 m below sea level.

optical stimulated luminescence (OSL): *see* p. 148.

orbit of the earth: the path which the earth uses around the sun (*see* Fig. 1–26).

Ordovician: *see* p. 35.

ore: a solid, naturally occurring mineral deposit of economic interest.

orogenic belts: mountain belts where orogeny occurs or has occurred.

orogenic periods: periods when mountains are built. In the earth's history there have been several important orogenic periods, of which the three youngest are best known: the Alpine orogeny (Tertiary in age), the Hercynian orogeny (Carbon-Perm), and the Caledonian orogeny (late-Silur/early-Devon) (*see* Fig. 1–31).

orogeny: the complex of processes by which mountains are built, including folding, faulting, thrusting, intrusion, volcanic activity etc.

orographic rainfall: precipitation caused by the uplift of air against a mountain barrier.

Ostracoda: small, bi-valved invertebrates, some of which live in salt water and some in fresh water.

outcrop: exposed part of a bedrock or of sediment beds.

outlet glacier: an ice tongue or arm extending out from a larger ice cover, an ice sheet, a plateau glacier etc. The tongue/arm can occupy a valley or a fjord.

outwash: sediments deposited by a glacial river beyond the front of the glacier.

outwash plain: the plain of outwash formed by glacial rivers beyond the ice front. The Icelandic term "sandur" is also used.

overloaded stream (river): a stream (river) which carries a heavier load than it can transport, and is therefore aggrading. This is typical for many glacial rivers.

overthrust: *see* tectonic structures.

overturned fold: *see* tectonic structures.

oxbow lake: a lake in an abandoned meander bend.

oxygen-isotope ratio: ^{16}O and ^{18}O are the two most commonly occurring oxygen isotopes in nature, and the $^{18}O/^{16}O$ ratio in the air is approximately 1/500. However, the ratio in which the two isotopes enter chemical compounds is temperature-dependent. Therefore, the isotope ratios in, for instance, speleothems, lacustrine carbonate sediments, and glacier ice (Figs. 1–20B, 1–21) have been used to record former temperature variations. The oxygen-isotope ratio in the ocean water and marine organisms varies with the amount of water stored in the late Cenozoic ice sheets also, and in some cases it can be difficult to distinguish the temperature isotope-signal from the glacial isotope-signal. The important deep-sea oxygen-isotope stratigraphy is based on the glacial signal (*see* p. 21).

Mass spectrographs are used to measure the isotope ratios, and the results are presented as $\delta^{18}O(‰)$, which is the deviation from a standard value. In dealing with carbonate shells, the isotope ratio in belemnite shells, the PDB, is used as a standard, and in dealing with water etc., the ratio in a Standard Mean Ocean Water (SMOW) is used as a standard.

ozone: O_3, a form of oxygen. The normal oxygen is O_2.

P

pack ice: an area of floating sea ice (broken pieces) which is not attached to the shore.

paha topography: an undulating surface of loess ridges and hills, which may contain some sand. The origin of the paha topography in the Midwest of USA has been much debated, and both an eolian origin and a fluvial-erosion origin have been suggested.

paleo-: old or ancient, used in combination with other words.

Paleocene: the oldest Cenozoic epoch (*see* Fig. 1–31).

Paleogene: The earliest part of the Cenozoic, before the Miocene (*see* Neogene).

Paleolithic: the old Stone Age (*see* p. 96).

paleomagnetic pole: an old position of the magnetic pole determined by measuring the direction of the remanent magnetism in rocks or sediments (*see* stratigraphy).

Paleontology: the study of fossils found in rocks and sediments.

paleosol: a buried old soil.

Paleozoic: *see* Fig. 1–31.

palsar: *see* frost features.

palynology: the study of pollen and spores in sediments to determine vegetation patterns and changes in vegetation over time (*see* pollen diagram).

palynomorfs: a term used about acid-insoluble micro-fossils like pollen, spores, dinoflagellates etc.

Pangaea: a Paleozoic supercontinent where all of the present continents were linked together.

parautochthonous rocks and sediments: nearly autochthonous rocks/sediments which have been moved a very short distance from their original position.

parent material: in soil science mostly used about the original material from which the material in the A_2 and B horizons have developed, corresponding with the material found in the C horizon.

parent rock: rock from which clasts, including erratics, have been derived.

patterned ground: *see* frost features.

peat: a brown to black organic residuum formed by the partial decomposition of dead plants and trees which have accumulated in wet places like bogs and marshes.

pebble size: *see* grain size.

pediment: a nearly horizontal erosion surface, veneered with detritus, usually spreading out from an erosion scarp.

pedology: soil science.

pelagic organisms: marine organisms which live in open ocean water, either as floating forms (planktonic) or as free-swimming forms (nektonic).

pelagic sediments: deep-sea sediments, as opposed to sediments derived directly from land.

pelite: a mudstone.

peneplain: a nearly flat surface formed near sea level by weathering and erosion over long periods, the final stage of old age in an erosion cycle, according to Davies. (*See* Davies' erosion cycle.)

penetration echo sounder: an echo sounder with strong signals which penetrate

and record beds, mainly soft beds, below the sea floor.

Pennsylvanian: *see* Fig. 1–31.

penultimate glaciation: the second-last glaciation.

Peorian loess: A Wisconsin-age loess in the Midwest of USA.

perched boulder: a large erratic in an unstable position on a rock surface, frequently carried by three or more small erratics which act as unstable legs (*see* Fig. 1–13).

periglacial area: an area near a glacier, dominated by cold climate and cold-climate processes. However, the term has been used in a wider sense also for areas with cold climate, permafrost etc., which are not related to glaciers.

perihelion: the point on the earth's orbit which is closest to the sun.

permafrost: *see* frost features.

permeability: the ability of rocks and sediments to transmit a fluid through pores and their interconnections (*see* geotechnical parameters).

Permian: *see* p. 35.

petrography: the part of geology dealing with description, classification and examination of rocks, particularly the igneous and the metamorphic rocks (*see* petrology).

petroleum: a naturally occurring complex of liquid hydrocarbons.

petrology: *see* petrography. Petrology has a broader scope than petrography and deals more with the origin and history of the rocks.

p-form or plastic form: *see* glacial-erosion features.

phenocrystals or phenocrysts: relatively large crystals scattered in the fine-grained matrix of a lava.

Phi-scale: a scale for the particle sizes of clasts based on the negative logs (base 2) of the particle diameters (*see* grain sizes).

pH-value: a quantitative expression of the acidity of a matter (water, soil etc.). Acid samples have pH values below 7 and alkaline samples have pH values above 7.

phyllite: *see* rock.

phyllum: *see* taxonomy.

piedmont glacier: the flat, distal part of one or several mountain glaciers which leave their valleys and spread out and coalesce on a flat surface at the foot of the mountains.

pingo: *see* frost features.

pipette method: a method used in grain-size analysis. The fine-grained sample (silt-clay) is dispersed in a fluid, and a pipette is used to collect samples at certain time intervals. The samples are then dried and weighed. The method is based on Stokes' formula (*see* p. 129).

piracy: when an actively eroding stream enlarges its drainage area by capturing parts of other streams, it is called piracy.

placer: a deposit, usually of sand or gravel, which contains valuable minerals such as gold.

planar bedding: *see* sediment structures.

planar cross-stratification: *see* sediment structures.

planation: formation of a flat surface, either by erosion or by deposition, or both.

plankton: a floating, small organism.

plasticity limit and index: *see* geotechnical parameters.

plateau glacier: a glacier on a relatively flat plateau (*see* Fig. 3–8).

plate tectonics: the movement of the earth's plates, which are large, rigid units of the earth's crust. The largest plates are continents. Where the thin oceanic plates collide with the thick continental plates, the margin of the oceanic plates can be forced down in a subduction zone. The world's deepest ocean basins (trenches) are formed over subduction zones. The margin of the continental plates and the sediments in the marginal zone between the oceanic and continental plates can be folded up on the landward side of the subduction zone. Molten lava which moves up through fissures, faults etc. in the ocean crust may form volcanic island arcs, and lava which moves up through the continental crust may form volcanic continental arcs. Zones where collision takes place between oceanic and continental plates or between two continental plates are the world's orogenic belts, where mountain chains are formed.

ocean-floor spreading: the drifting apart of the oceanic crust on both sides of a rift in the crust. Oceanic ridges and new crust are formed by volcanic activity in this kind of rift. The Mid-Atlantic ridge, on which Iceland is located, is an oceanic ridge.

sea-floor spreading: *see* ocean-floor spreading, above.

playa: the flat central part of an enclosed wide basin where shallow lakes are formed periodically.

Pleistocene: *see* p. 35.

Pleniglacial: the Weichselian Glaciation, excluding the relatively warm early and late parts, the Amersfoort, Brörup, Bölling, and Alleröd interstadials. The term is not much used.

Pliocene: *see* Fig. 1–31.

plough marks: *see* glacial-erosion features.

plucking: *see* glacial-erosion features.

plunge: the inclination of, for instance, an axis (*see* tectonic structures).

plutonic rock: a rock formed at considerable depth below the earth's surface by crystallization of a magma or by chemical alteration.

pluvial lake: a former lake formed by increased rainfall in an area which today is a dry desert.

pluvials: periods with increased rainfall in areas were pluvial lakes were formed.

pockmarks: shallow, small, generally circular depressions on the sea floor formed where gas from the sediments in deeper-lying beds escaped.

podsol: *see* soil.

point bar: the bar of sand and/or gravel which accumulates on the depositional side of a meander bend. The sediments in a point bar usually show an upward fining sequence.

Polar Front: the zone where cold air from the polar regions meets warm, moist air from lower latitudes. The term has been used in marine geology also, for the zone where cold polar surface water meets warmer subpolar water, a zone which approximately corresponds with the limit of winter ice cover on the polar oceans, and the 2°C water-temperature isotherm (*see* Arctic Convergence).

polar glacier: a glacier whose subsurface temperatures in the accumulation area are below freezing throughout the year, although the temperature may be at the pressure melting point at the base of thick polar glaciers.

pollen: small grains which are reproductive units of seed plants generally spread in large quantities by wind or, in smaller quantities, by insects. The fossil pollen consists entirely of the durable outer crust, the exine (*see* palynology and pollen diagram).

pollen diagram: a diagram which presents the changes in vegetation observed in a

stratigraphic section by pollen analysis (*see* Fig. 2–68).

the percentage pollen diagram: based on recorded percentages of counted pollen grains of the different plant taxa within each analyzed sample.

the concentration pollen diagram: records the changes in the total number of pollen grains of each taxon per unit sediment volume.

Other kinds of pollen diagrams are used also, such as the pollen influx diagram, which records the number of pollen grains of each taxon deposited on 1 cm^2 per year. The sedimentation rate must be known to construct this diagram.

In interpreting a pollen diagram note that: (1) Some taxa produce much more pollen than others; for instance, Pinus, Betula, Alnus, and Corylus are heavy pollen producers. (2) The percentage of long-distance-transported (by air) pollen is frequently high in samples from areas with low pollen production. (3) Redeposited (secondary) pollen grains are common in some sediments, and can, in some cases, be hard to distinguish from the primary pollen grains. (4) A pollen grain can usually be identified to family or genus level, and in some cases also to species level.

The following are some of the most commonly occurring plant groups in pollen diagrams: arboreal pollen (from trees), non-arboreal pollen (from herbs), and pollen from shrubs and dwarf shrubs. The non-aboreal pollen dominates in areas with open vegetation, such as above and north of the forest limit. The following are some trees which occur mainly in the Subarctic (A), the Boreal (B) and in the Broad-leaf (C) forest zones: *Betual* (birch, A+B), *Picea* (spruce, A+B), *Pinus* (pine, A+B) *Larix* (larch, A+B), *Alnus* (alder, A+B+C), *Abies* (fir, A+B), *Ulmus* (elm, C), *Quercus* (oak, C), *Fagus* (beech, C), *Fraximus* (ash, C), *Tilia* (lime, C), *Corylus* (hazel, C), *Carpinus* (hornbeam, C), *Taxus* (yew tree, C).

Note that there are classifications which are slightly different from the one used above.

The following are the names of some non-arboreal and shrub taxa commonly occurring in pollen diagrams: *Salix* (willow), *Juniperus* (juniper), *Ericaceae* (heather), *Hedera* (ivy), *Hippophae* (sea buckthorn), *Ilex* (holly). Herbs: *Graminaeae* (grasses), *Cyperaceae* (sedges), *Artemisia* (wormwood, mugwort), *Plantago* (plantain), *Rumex* (sorrel), *Dryas* (mountain avens).

population: a term used in biostratigraphy for a group of plants or animals found in a sediment or rock unit.

pore: for sediments the term means a space between the clasts which is not filled with mineral matter.

porosity: the ratio between pore space and the total volume of a rock or a sediment (*see* geotechnical parameters).

porphyry: a rock containing conspicuous phenocrysts in a fine-grained matrix.

Postglacial: the term Postglacial has been used synonymously with the term Holocene. However, postglacial (without a capital P) is more commonly used in a wider sense for the time following the deglaciation of an area.

pothole: the term is most used about a pot-like depression in bedrock, frequently several meters deep and generally less than 3 m wide, formed at the base of a waterfall by stones that spin around and abrade in the pot. Potholes are frequently formed at the base of waterfalls in glacial crevasses (*see* Fig. 3–20).

Preboreal: *see* Fig. 2–68.

pressure melting point: the temperature at which ice melts at the existing pressure. The pressure below a glacier varies and so does the pressure melting point, which drops with increasing pressure.

primates: the order of the class *Mammalia* which includes *Homo sapiens*.

proglacial features: features related to glaciers, but formed beyond the glacier itself.

prograadation: the seaward advance of the shore caused by deposition of sediments transported by rivers etc.

protalus rampart: a ridge of rubble/debris which has accumulated at the foot of a snowbank, generally at the foot of a talus slope.

Protozoa: single-celled organisms belonging to the phylum *Protozoa*, such as foraminifera, radiolaria and dinoflagellates.

proximal part or side: in sedimentology, a term used about the upstream part of a terrace (delta plain), or the ice-contact side of an end-moraine ridge etc. The opposite of the distal part or side.

pseudo-cross-bedding: *see* sediment structures.

pseudo-fossil: a rock object or a pattern on a rock surface which resembles a fossil, but is not a true fossil.

pumice: a light, very vesicular lava, generally with rhyolitic composition.

push moraine: an end moraine formed by the bulldozing of an advancing ice front; a ridge pushed up by the ice front (*see* moraine).

pyknocline (pycnocline): the change in density of a fluid in one direction, for instance in a layer of water with a vertical density gradient.

pyroclastic: volcanic material that has been aerially ejected. Pyroclastic rocks consist of pyroclastic material.

pyroxen: *see* minerals.

Q

quartz: a mineral with the chemical composition SiO_2. One of the most common minerals in rocks, it occurs in several varieties.

quartzite: *see* rock.

Quaternary: *see* p. 36.

Quaternary geology (Qg): the geology of the Quaternary Period, about the last 1.7 million years, or the geology of the last 2.5 million years. Qg is traditionally focussed on climatic changes and the character, origin, and history of the landscape, and of the sediments which lie on top of the bedrock. Endogene processes, features, and rocks are generally omitted. In Qg, scientists from most geological fields (like glacial geology, sedimentology, geomorphology, lithostratigraphy, biostratigraphy, mineralogy, marine geology, etc.) cooperate with climatologists, archeologists, botanists, zooligists etc. and synthesize the results. A most important topic in Qg is the late Cenozoic climatic changes and corresponding environmental variations, such as variations in sea level, glacier fluctuations, vegetation, ocean currents etc. A main theme is the history of pluvials and interpluvials, and glacials and interglacials. It was generally believed that the last global cold period with large glacials and interglacials lay entirely within the Quaternary Period. However, results of recent research indicate that this period, with large glacial expansion, started about 2.5 million years ago, which is within the late part of the Pliocene. Therefore, the geological history of the last 2.5 million years has been, and will most likely be, a main theme in Quaternary geology.

Quaternary research: includes research within many scientific branches used in working out the history of the Quaternary Period (or the last 2.5 million years), and used in recording the properties of the Quaternary sediments and features.

quick clay: *see* p. 129.

R

radioactive isotopes: unstable atoms that change into other atoms by emitting charged particles.

radiocarbon dating: *see* radiometric dating.

radio echo sounding: a method used to measure thicknesses of ice, for instance, by means of recorded travel time for elctromagnetic (radio) pulses.

Radiolaria: a planktonic, microscopic marine *Protozoa* with siliceous skeleton.

radiometric dating: dating of rocks, minerals, sediments, and fossils by means of radioactive isotopes. With the introduction of Accelerator Mass Spectrometry (AMS) and the recent improvements of the AMS technique and the techniques of other mass spectrometers, a great number of radioactive elements which occur in very small quantities can now be quantitatively determined. This opens the possibility to use many different radioactive isotopes, like ^{36}Cl, ^{26}Al, ^{10}Be, ^{32}Si, ^{226}Ra, ^{39}Ar, ^{3}He, ^{21}Ne, in dating Quaternary deposits. However, the methods in which these isotopes are involved are still being developed and are generally not commonly used. The dating methods most used are based on the radioactive isotopes ^{14}C, ^{238}U, ^{235}U, ^{230}Th, ^{40}K, ^{39}Ar, ^{210}Pb, and they will be mentioned in the following:

Uranium-series (^{238}U, ^{234}U, ^{230}Th) *dates:* The three elements decay to lead isotopes. ^{230}Thorium was formerly called Ionium. The dating range is from about 1000 to about 350 000 years B.P. Corals, speleothems, lacustrine marl, peat,

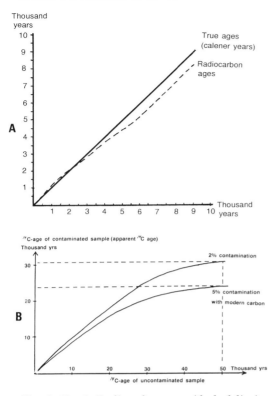

Fig. 4–13. A: *Radiocarbon ages (dashed line) compared with true ages (heavy line) of organic samples younger than 9000 years. Note that there is almost no age difference for samples younger than 3000 years, and for older samples the radiocarbon ages are always younger than the true ages. The lines are based on information presented by Stuiver, M. and Reimer, P.J., 1986.*

B: *Graphs showing the influence of contamination with modern carbon on the observed ¹⁴C age. Note that 50 000 year old samples which are 2% and 5% contaminated have apparent ¹⁴C ages of about 30 000 years and about 24 000 years, respectively.*

bone, and molluscs have been dated. However, if the dated sample lay in an open system with free chemical exchange with, for instance, the ground water, the date can be incorrect. This is true for many dates of molluscs.

Radiocarbon (¹⁴C) dates: There are three isotopes of carbon occurring in nature, ¹²C, ¹³C and ¹⁴C. The percentages of the three in the atmosphere are about 98.9%, 1.1% and 0.000000000118%, respectively, and they are all oxidized to carbon dioxide (CO_2). ¹⁴C is formed in the upper atmosphere by cosmic-ray bombarment of nitrogen atoms (¹⁴N) (*see* p. 19). The CO_2 in the atmosphere is

transformed by photosynthesis to organic matter in plants on which animals feed. Therefore, all living organic matter contains carbon isotopes in about the same ratio as in the air. When the plant or animal dies the dead organism receives no more carbon; the radioactive ¹⁴C in the organism starts to decay, and the radioactive clock starts. The disintegration takes place at a constant speed, and Libby calculated that half of the original ¹⁴C disintegrates in 5568 ± 30 years. Later the half-life of ¹⁴C was more accurately determined at 5730 ± 40 years. However, by international agreement all radiocarbon dates are based on Libby's value. The ages are presented as years B.P., and the year 1950 is used as present. Radiocarbon dates from about 100 years to about 70 000 years have been presented, but most laboratories have an upper limit at 40 000–50 000 years B.P. Dates above 60 000 are obtained by enrichment of ¹⁴C.

The traditional ¹⁴C dating method is the proportional gas-counting method, which is used by most ¹⁴C dating laboratories. The organic carbon is transformed to CO_2 gas (burned), and the gas is brought into a chamber which is surrounded by counters where the radioactive radiation is counted. The system is shielded against cosmic radiation. A few laboratories use the liquid scintillation method, in which the carbon is dissolved in a liquid when the counting is done.

Accelerator dating (AMS dating) of carbon is a new method in which the content of ¹⁴C is measured in accelerator mass spectrometers. This method allows the dating of very small samples, and it is increasingly used today.

Problems: There are several problems involved in using the ¹⁴C method to date fossil organic matter, and the most important will be mentioned briefly in the following:

Contamination. The most serious problem in obtaining correct dates is often a contamination of the organic matter which took place before the sample was collected, in particular contamination with younger carbon (for example, roots that penetrate the older sediments, dissolved humic acids transported with the ground water, and ionic

exchange in carbonates). The importance of this contamination increases with the age of the fossil matter, since the amount of the original ¹⁴C decreases drastically with age (it is reduced to 1/16th in a 23 000 year old sample), and a very small contamination results in a too-young age (*see* Fig. 4–13). Good sealing quality of the host sediment is an improtant factor to avoid contamination. Many scientists still consider most of the very old radiocarbon dates to be minimum dates.

Ages that are too old are common in lake sediments in limestone areas, because "infinite old" carbon from the limestone is dissolved in the water, and plankton and water plants use this old carbon together with the "normal" carbon. This is called the hard-water effect, and it is well known from hard-water lakes, where the ¹⁴C dates are frequently 1000–2000 years too old.

Changes in atmospheric ¹⁴C content. The ¹⁴C dating method is based on the assumption that the content of ¹⁴C in the air has been constant. However, comparisons between tree-ring dates and ¹⁴C dates of the same tree-rings show that the ¹⁴C dates depart (with increasing ages) from the tree-ring dates, and from 6000 to 9000 years B.P. the ¹⁴C dates are about 700 years too low (*see* Fig. 4–13). This indicates that the atmospheric content of ¹⁴C about 6000–9000 years B.P. was higher than today.

Comparison between high-precision, mass-spectrometer, U-series dates and ¹⁴C dates of corals, back to 30 000 years ago, show that the ¹⁴C ages are systematically too low, with a maximum value of about 3500 years at 20 000 years ago. It is suggested that changes in cosmic nuclide production, linked with a decrease of the earth's magnetic field, is the main cause for the changes in atmospheric radiocarbon.

Plateau levels. Recent studies of radiocarbon-dated annual lacustrine laminae show that the radiocarbon age is the same for all laminae deposited during certain periods, approximately 200 years long. One period apparently occurred just before 12 000 years B.P., and another shortly after 10 000 years B.P.

Fractionation. A slight isotopic fractionation takes place when CO_2 is absorbed in ocean water during photosynthesis and other processes. The radiocarbon laboratories correct for isotopic fraction by measuring the ^{13}C content by means of mass spectrometers, and normalize the value to $\delta^{13}C$ = minus 25 o/oo PDB, which is the average value for terrestrial plants.

The reservoir effect. The mixing of water in the oceans is usually slow, and deep water can be several thousand years old in some areas. The ^{14}C age of animals living in this kind of "old" water is too old. For instance, in the Ross Sea area of Antarctica, some living animals have a ^{14}C age of 1000–1500 years, which is called the reservoir age at this locality. Most surface ocean water, for instance along the coasts of North America and northern Europe, has a reservoir age of about 400–500 years, which should be subtracted from the obtained ^{14}C age of fossils of marine organisms from this water. A reservoir effect has even been observed in many fresh-water lakes that are not hard-water lakes, and organisms living in those lakes are frequently slightly too old.

Potassium-argon (^{40}K–^{40}Ar) dates: The transformation of ^{40}K to ^{40}Ar in volcanic rocks and ash has been used to date old Quaternary deposits, generally older than 1 million years.

Argon-argon ($^{40}Ar/^{39}Ar$) dates: Using a high-precision laser, Mass Spectrometer technique, this method has been used to date volcanic samples considerably younger than 1 million years.

Lead (^{210}Pb) dates: Radioactive lead (^{210}Pb) has a half-life of 22.26 years, and it has been used to date sediments which are younger than 150–200 years.

raised bog: a bog with a convex surface.

ravine: a V-shaped, generally small and short valley or gully eroded by water. Ravines are common on ground which is capped by relatively impermeable soft rocks or sediments such as clays, where most of the water runoff takes place on the surface (*see* Fig. 3–14).

recessional moraine: an end moraine formed during a temporary stop in the retreat of an ice front, or during a slow-down of the retreat. The term has even been used for moraines formed during small readvances of an ice front during a period of general retreat.

recent: a term used in a broad sense for a period close to the present day. The term has no accepted definition, but it has in some cases been used for the Holocene.

recumbent fold: *see* tectonic structures.

red clay: a fine-grained sediment found on ocean floors at depths below 5000 m.

reflection seismology: a method used to record rock beds below the earth's surface, or ocean floor, based on the reflection of seismic waves from discontinuities between beds. The waves are produced by explosives on the earth's surface or ocean surface.

refraction seismology: a method used to record rock beds below the earth's surface, or ocean floor, based on the refraction of seismic waves which are generated on the earth's (or ocean) surface.

refugium: an area where some plants or animals survived an unfavorable period, generally a glaciation, when plants or animals of the same kind were extinguished in surrounding areas.

regelation: the refreezing to ice of water within glacier ice at the pressure melting point.

regression: the change to dry land of a sea-covered near-shore zone, resulting from a drop in shore level caused by a eustatic drop in sea level or an isostatic or tectonic uplift of the land.

rejuvenation: a term used in geomorphology about the renewed erosion which takes place in an area, generally caused by tectonic uplift or by a drop in local base level.

relative age: an age determined by comparing beds (zones) in different stratigraphic sections, and not obtained by exact dating (*see* geochronology).

remanent magnetism: lasting magnetism, such as paleomagnetism in rocks and sediments.

repose angle or angle of repose: the maximum angle at which the surface of a deposit can rest under certain physical conditions.

residual deposit: the material remaining in place after decomposition of a rock.

reverse fault: *see* tectonic structures.

reworked sediment: a sediment which has been transported from its original place of deposition, and has been more or less transformed on its way.

rhyolite: a fine-grained equivalent of granite (*see* rock).

rhythmite: *see* sediment structures.

ria coast: a coast with bays/sea arms that are drowned, stream-eroded valleys which have not been glaciated.

Richter scale: a scale used to measure the magnitude of earthquakes. The scale starts near zero and about 9 is the strongest recorded quake.

rift valley: *see* tectonic structures.

rift zone: a zone with large-scale earth-crustal fractures and faults.

rill mark: *see* sediment structures.

ripple: *see* sediment structures.

Riss Glaciation: *see* Fig. 2–1.

roche moutonnée: *see* glacial-erosion features.

rock: a solid material formed naturally at or below the earth's surface. There are three classes of rocks: igneous, sedimentary, and metamorphic. The igneous rocks consist of minerals, the sedimentary rocks of clasts and/or minerals, and the metamorphic rocks of minerals resulting from alteration of clasts and/or minerals of an original rock which was subjected to heat and/or pressure, generally caused by tectonism. There are many different kinds of rocks within each rock class, and only a few of the most common can be mentioned:

Igneous rocks are formed from magmas and there are two main groups:

1. *Lavas* are formed at the earth's surface by rapid cooling of the magma. They are therefore fine-grained, but they may have scattered larger crystals called phenocrysts:

basalt: a blackish, fine-grained basic rock of mainly calcic plagioclase and pyroxene.

rhyolite: a light-colored acid rock, frequently glassy, with a chemical composition similar to granite. A glassy type is called obsidian.

porphyries: various kinds; all have visible, scattered larger crystals, phenocrysts, as for instance the well-known rhombporphyry (*see* Fig. 2–4).

2. *Plutonic rocks* were formed at various depths below the earth's surface. They were slowly solidified, and that allowed coarse mineral grains to develop. They are therefore coarse-grained.

Granites are some of the most common acid rocks. They have a silica content of about 70% and contain grains of quartz, alkali felspar, and mica.

Syenites are intermediate, light-colored rocks with grains of mainly alkali felspar and some darker grains, mainly of hornblende or biotite.

Diorites are intermediate, somewhat darker-colored rocks with oligoclase-andesine felspar, hornblende, biotite or other mafic minerals, with less than 10% quartz, and some alkali felspar.

Gabbros are dark, basic rocks composed of calcic plagioclase and pyroxene.

Metamorphic rocks:

Gneiss is a coarse-grained, banded rock, commonly rich in felspar.

Schists are fine- to medium-grained rocks which have foliation surfaces and are usually named after the dominating mineral, such as mica-schist.

Phyllite is a foliated, lower-grade metamorphosed rock than mica-schist, and *slate* is a still lower-grade metamorphosed rock than phyllite; all three originate from claystones.

Quartzites consist almost solely of quartz, and they are metamorphosed sandstones or siliceous ooze.

Sedimentary rocks are formed by petrifaction of sediments, including sediments resulting from precipitation from solutions or secretion from organisms. The most common are: claystone or mudstone, siltstone, sandstone, greywacke (an impure, dark sandstone). Conglomerate, breccia, tillite, etc. are sedimentary rocks dominated by clasts. Limestone consists of more than 50% $CaCO_3$, frequently of cemented shells, but in some cases of chemical precipitates. Chalk is a very pure white limestone. Dolomite consists of $CaMg(CO_3)_2$ and is generally formed from $CaCO_3$ by adding Mg. (*See* also flint and chert.)

rock glacier: a glacier-like lobe of angular rock debris with some interstitial ice, which moves slowly downhill. The rock glaciers are usually fed from steep head-walls or mountain slopes where large quantities of rock material are quarried by frost action, and they generally lie in a zone immediately below the zone of ordinary cirque glaciers.

Rogen moraines: relatively closely spaced moraine ridges oriented transversely to the observed former ice-flow direction. The ridges generally lie on wide valley floors, and they form irregular patterns. However, they are not lobate, and are therefore not interpreted to be end moraines. According to most of the many suggested interpretations, they are formed subglacially. Drumlin features are usually related to the moraines.

rose diagram: a diagram used to show the orientation of the long axes of clast etc. (*see* fabric analysis).

roundness: the roundness of a clast is usually determined according to scales ranging from angular to well rounded (*see* p. 153). Other, more elaborate methods have been developed to describe the shape or sphericity of clasts.

S

Saalian (Saale) Glaciation: *see* Fig. 2–1.

saltation: a mode of transportation of clasts along the river bed by bouncing.

salt dome: a diapir consisting of salt (*see* tectonic structures).

salt marsh: a marsh which is periodically flooded with salt ocean water.

sand size: *see* grain size.

sandstone: *see* rock.

sandur: *see* outwash plain.

sandur delta: *see* glaciofluvial delta.

Sangamon Interglacial: *see* Fig. 2–1.

sapropel: an aquatic ooze rich in organic matter.

sastrugi: a streamlined, sharp-edged, low ridge of hard-packed snow on a snow surface, formed by strong winds. The sastrugi is commonly combined with shallow, elongated depressions which are formed by wind erosion.

savanna: the grassland which lies between the tropical rainforest and the desert belts.

scarp or escarpment: a steep slope or cliff along the margin of a flat surface, plateau or terrace.

schist: *see* rock.

schistocity: the arrangement in (nearly) parallel layers of minerals in metamorphic rocks.

Schmidt net: *see* fabric analysis.

schotter: a German term used for stream-transported, sorted gravel or a body of gravel. Decken-schotter is a cover of gravel and Vorstoss-schotter is a unit of gravel formed during a glacial advance.

scree: a steep sheet of rock debris below a steep cliff or rock face (*see* talus).

sea-floor spreading: *see* plate tectonics.

section: generally used about a vertical or nearly vertical exposure of rock beds or sediment beds.

sediment: material settled from a solution or a suspension, or material that has been transported by water, wind, ice or gravity, and has come to rest.

sedimentary facies: *see* facies.

sedimentary rocks: *see* rock.

sedimentology: the branch of geology dealing with the sediments and sedimentary rocks, their character and origin.

sediment structures:

angular disconformity: see unconformity (p. 188).

bed: in sedimentology, used when describing a well-defined strata or layer in a stratigraphic section, or about the lowest grade unit in lithostratigraphic nomenclature (*see* stratigraphy). Layers thinner than 1 cm are usually called laminae.

bedded deposit: a stratified deposit with well-defined layers; it may contain some laminae.

bioturbation: disturbance, including burrowing of sediments caused by the activity of animals which live in the sediments: *see* trace fossil (p. 188).

boudinage structure: a pillow-like structure caused by cross-fractures which divided the bed into segments that developed into "pillows". In section, the structure may resemble a string of linked sausages.

bouma sequence: the most characteristic stratification of a density-current deposit (*see* Fig. 4–1).

burrow: a "cylindrical" tube filled with clay, silt or sand made by animals that lived in the mud/sand. It is evidence of bioturbation.

clastic dyke (dike): a crack, fissure etc. filled with clastic material (sand, silt, clay) which is generally different from the material on the sides of the dike walls.

collapse structures: a term used in sedimentology about structures like folds and faults resulting from the collapse of stratified sediment units which rested on, or were supported by, glacier ice that subsequently melted.

concordant: concordant beds are parallel beds (*see* discordant, p. 187). The term is used both for igneous and sedimentary beds. Concordant mountain summits lie on a uniform surface.

conformable: a term used when a bed, or a unit of beds, lies parallel with the surface or bed on which it rests.

convolute lamination (bedding): a unit of strongly contorted laminae (beds) within a succession of parallel beds.

cross-bedding: a main bed containing short parallel beds which lie obliquely to the main surface of deposition represented by the main bedding planes (*see* trough cross-bedding, and planar cross-stratification, below).

cross-lamination: lamination with laminae oriented obliquely to the main bedding planes. The terms cross-lamination, cross-bedding, and cross-stratification have sometimes been used synonymously.

cross-stratification: a general term for cross-bedding and cross-lamination.

current bedding (lamination): the same as cross-bedding (lamination) or cross-stratification caused by currents.

current ripple: a ripple caused by a current (*see* ripple).

dip: the angle between a surface or bed and the horizontal.

disconformity: see unconformity (p. 188).

discordance: lack of parallelism; the opposite is concordance.

discordant: a term used when a surface, a bed or unit parallel beds are cut by a surface or bed with a different dip.

flute mark (cast): see sole marks, below.

graded bed: a bed which shows a gradation in grain size, generally from coarse below to finer above. However, beds with inverse (reverse) grading (from fine to coarser grains) do occur.

hiatus: a surface in a succession of sediment strata which represents a relatively long break in sedimentation.

horizon: a term usually used about a thin bed or a thin zone within a bed, with characteristic lithology or fossils.

ice-wedge cast: the wedge-shaped sediment-filling in a wedge which was formerly filled with ice (*see* frost features).

imbricated clast: clasts with their flat surfaces dipping, or longest axes plunging upstream in river beds, upglacier in till beds etc.; a sort of shingle-on-a-roof orientation of the clasts.

interbedded: beds which lie between, and parallel with, beds of a different kind in a stratified unit. The term interstratified is also used.

inverted: beds which are tectonically turned upside down.

involution: contortion of surface material (soil etc.) by various processes like creep, slumping, loading, and expulsion of water – processes that are common in periglacial areas. The term is also used in structural geology about some kind of refolding of two nappes.

lamina (plural: laminae): traditionally this term has been used for a layer that is no more than 1 cm thick. However, in some textbooks the term has been used in a wider sense, and it has been suggested to use the term "thick lamina" for a 1–3 cm thick bed, and "very thick lamina" for a bed thicker than 3 cm.

lamination: a layering/bedding where the layers are laminae, or where most of the layers are laminae.

laminite: a deposit consisting of mainly laminae.

lenticular: lens-shaped.

limb: used in connection with a fold (*see* tectonic structures).

load cast: a flame-like (cumulus-cloud-like) structure in a unit of fine-grained sediments (clay, silt or fine sand) caused by the load of overlying sediments at a time when the fine-grained sediments were saturated with water under hydrostatic pressure.

mega ripple: see ripple.

offlap sequence: beds with gradually smaller lateral extension in an upward sequence, deposited in a gradually shrinking sea or lake. The opposite of an onlap sequence.

onlap: beds with increasing lateral extension in an upward sequence, deposited in a rising sea or lake.

planar bedding (tabular bedding): the bedding in a unit with flat-lying parallel beds.

planar (tabular) cross-stratification: a planar bedded unit with cross-bedding or cross-lamination within one or several of the beds.

pseudo-cross-bedding: bedding produced by the migration of ripples, and resembling cross-bedding. Most scientists call this kind of bedding ripple cross-bedding.

rill mark: a shallow, dendritic, erosional structure generally formed by currents in a thin water layer that flows on a sediment surface.

ripple: a ridge, usually of sand, formed

by wind, or water currents or waves. The ripples lie transversely to the current- and wave-motion direction.

Small ripples are generally 0.3–6 cm high and have about a 4–60 cm wave length (distance between two successive crests).

Mega ripples are 60 cm–1.5 m high and have a 30–1000 m wave length.

wave ripples: usually symmetrical, but they can be asymmetrical. The crests are usually straight, frequently show bifurcuation, and are usually pointed.

current ripples: usually asymmetrical, but they can be symmetrical. They frequently have foreset lamination, and they usually migrate downstream, but may in some cases migrate upstream.

wind ripples: frequently have long, straight crests. They are asymmetrical, generally with no visible lamination, and they have a concentration of larger grains at the crests.

Different kinds of parameters, such as symmetry parameters, have been used in identifying the different kinds of ripples. See textbooks, for instance Reineck and Singh, which contains 37 pages about ripple formation and classification.

rhythmite: a pair of two laminae or thin beds in a sequence of rhythmites. The pair frequently consists of a light and a darker laminae/bed, which can represent the sedimentation through a summer and a winter season, or through seasons of shorter or longer duration. A varv is a rhythmite (*see* varv, p. 188).

set: a group of parallel or nearly parallel features like beds, dikes, veins etc., for instance topset beds, bottomset beds, and foreset beds in a delta.

sole marks: structures visible on the lower bedding-plane surface formed by erosion of flowing water on this surface when the sediment was deposited. Sole marks are visible after the sediments have been petrified and the sole has been exposed by erosion. There are several different kinds of sole marks such as flute marks (casts), current crescents etc. Sole marks are used to denote which side was originally up when a bed was deposited.

stratification: a layering in rocks or sediments. The layers can be beds, laminae, lenses etc.

stratum: a term used both for a uniform

Shore zones

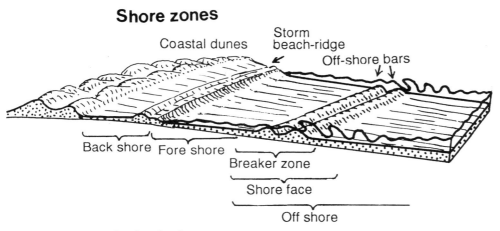

Fig. 4–14. Features related to the shore zone.

stratified unit consisting throughout of about the same kind of clasts, and for a single sedimentary layer regardless of thickness. Strata is plural of stratum.

tabular bedding: see planar bedding (p. 187).

trace fossil: structures on the sediment surfaces or within the sediments caused by biogenic activity, such as burrows and tracks formed by living animals (*see* bioturbation, p. 186).

trough cross-bedding: cross-bedded infills in channels etc., which look like troughs in vertical cross sections; a common phenomenon in river channels.

truncate: to cut (the end) off, for instance of a bed, a valley etc.

unconformity: an erosion surface separating a unit of younger strata from a unit of older. If strata in the young unit have a different dip than the strata of the older, the term angular unconformity is used.

varve: a rhythmic couplet which was deposited in one year (*see* varved clay).

wedge out: a term used to describe a bed or a unit of beds which become gradually thinner to a point where they end (wedge out).

sediment texture: *see* texture.

seismic: phenomena related to earthquake activity or to artificially induced vibrations of the ground produced in connection with the exploration of structures below the earth surface. (*See* reflection seismology, refraction seismology, and shallow seismic recording.)

seismograph: an instrument which records earthquakes.

seismology: the science dealing with seismic activity.

Series: a time-stratigraphic unit (*see* stratigraphy).

set: *see* sediment structures.

settling velocity: *see* Stokes' law, and grain size.

shallow seismic recording: recording of the upper sediment beds below the ocean floor, or on land, by means of low-energy vibration (wave) sources like air-gun, boumer, sparker, and penetration echo sounder.

shear plane: *see* tectonic structures.

shear strength: the internal strength or resistance of a rock or sediment against being sheared (*see* geotechnical parameters).

shear stress: a stress tending to make a rock or a sediment break and slide along a plane, a shear plane.

shelf: *see* continental shelf.

shore zones: *see* Fig. 4–14.

sial: the part of the earth's crust which is rich in silica and aluminum. It forms the continental crust (*see* sima).

significance level: *see* fabric analysis.

sill: a bed-shaped igneous or clastic intrusion which lies concordant with the host beds.

silt: *see* grain size.

Silurian: *see* p. 35.

sima: the part of the earth's crust which is rich in silica and magnesium. The sima is heavier than the sial, and it forms the ocean crusts.

sinter: a deposit of siliceous material derived from a hot spring (*see* travertine).

skewness: *see* grain-size parameters.

slate: *see* rock.

slickenside: a polished and striated surface developed by the friction during the formation of a shear plane.

slip: *see* tectonic structures.

slope wash: the downhill movement of material on a slope caused by gravity and washing by water, but not by rivers.

snow line: the lower limit of the winter snow cover observed in the following fall when this limit lies at its highest. The snow line is usually observed on glaciers where it corresponds roughly with the firn line and the equilibrium line. Today the term equilibrium line is most used. Note that other kinds of snow lines have been defined, such as the regional snow line, the orographic snow line, and the temporary snow line.

soil: the thin zone of unconsolidated material at the earth's surface which has been subjected to physical, chemical, and biological processes. Soils have A, B, and C horizons, and the A and B horizons develop differently in different climates and on different parent material (*see* p. 139).

soil samplers: samplers used to collect samples of subsurface sediments, generally from the upper 1 m. The term auger is frequently used for a soil sampler, which is drilled into the ground. There are many different kinds of samplers, such as: (1) A cork-screw shaped auger on which fragments of the sediments stick to the screw. (2) A box (bucket) auger which has a cylindric chamber that is drilled into the ground. Cuttings from the sediments are collected in the chamber. (3) One of the most used "soil samplers" is a 1–2 cm thick metal rod with grooves along the sides, in which the sediments stick and can be identified.

solar constant: the solar energy received per square cm, per minute, at a surface normal to the radiation direction immediately outside the earth's atmosphere, when the earth is at its mean distance from the sun.

solifluction: the slow movement downhill of saturated soil caused by gravity, mostly in permafrost areas where the movement is supported by frost action, but also in other areas where it is not supported by frost (*see* gelifluction and congelifluction).

sorted circles, nets and polygons: *see* frost features.

sorting: the process by which a sediment is being sorted. The term is also used as a descriptive term and means the degree of similarity or spread in grain sizes within a sediment (*see* grain-size parameters).

sparker: a seismic energy source used to produce waves that are reflected from reflectors in the sediments on the sea floor. The sparker waves penetrate deeper than the echo waves, frequently 100–400 m, depending upon the kind of sediments in the beds.

speleology: the scientific exploration of caverns.

speleothem: limestone deposit in a cave, such as a stalactite or stalagmite.

spit: a narrow, pointed accumulation, usually of sand, into the sea or a lake.

stade: the time corresponding with a glacial deposit or glacial phase.

stadial: *see* interstadial.

Stage: a time-stratigraphic unit next in rank below Series (*see* p. 190).

stagnant glacier: generally said about a glacier which is both climatically and dynamically dead, the opposite of an active glacier, although the term has been used also about glaciers which are climatically dead and still may be slightly active.

stalactite: a deposit of calcium carbonate, hanging down from the roof of a cave, frequently in the shape of a pointed, long "finger".

stalagmite: a column of calcium carbonate deposit which rises from the floor of a cave.

step fault: *see* tectonic structures.

stereoscope: an optical instrument used to see the terrain stereoscopically on sets of two photographs that are shot for that purpose.

Stokes' law: a law expressing the velocity by which a spherical particle (smaller than 0.2 mm) is dropping in a fluid: $v = Cr^2$, v = velocity, r = radius, C = a "constant" which depends upon the specific weight of the particle and the density of the fluid. Since most sediment particles are not spherical, in particular the clay particles, the calculated radius which corresponds to the observed drop speed is not the true radius of the particle, but the so-called "effective radius", the radius of a spherical particle, which drops with the same velocity as the observed particle.

stone polygon, stone ring, stone stripe: *see* p. 138.

storm beach: a beach ridge, generally of coarse clasts, formed above sea level during heavy storms, frequently between 1 m and 2 m above sea level.

stoss-sides: slopes which have been ice covered and were facing the glacier-flow direction. The term is used in connection with sand dunes etc. (*see* also glacial-erosion features).

strain: the change in size and shape of an object, for instance a rock, caused by a stress.

strand-flat: a plain which lies close to sea level, from somewhat below to some tens of meters above, and frequently ends in a sharp break at the foot of a cliff at its inner margin. The true strand-flat is developed mainly under Arctic conditions, and it can be several kilometers wide. Various processes took part in the formation, such as frost shattering and shore processes.

stratification: *see* sediment structures.

stratigraphic unit: a unit of rock or sediment layers which has a related lithology or fossil content.

stratigraphy: the part of geology dealing with the study of stratified rocks and sediments, their lithology, fossil content, depositional environment, age and correlation. In stratigraphy there are three main branches: lithostratigraphy, biostratigraphy, and chronostratigraphy. Some rules for nomenclature within the three branches are shown below.
Lithostratigraphy (rock stratigraphy): This is based on the lithology of the sediments and the interpretation of the environment in which the sediments were deposited. The various lithological elements such as grain size and sorting, rounding, morphology, orientation, and petrology/mineralogy of clasts, and textures like various types of stratification, folds, shear planes etc. are observed and used to interpret the sedimentary environment. This is in fact a main theme for the many textbooks in sedimentology, some of which are close to 1000 pages. In describing the stratification in a section, several informal terms like bedding, layering, stratification, layers, beds, strata, units, subunits etc. can be used. However, in giving a bed or a unit of beds a formal name, certain rules should be obeyed. They are presented in a stratigraphic guide written by Hedberg (1976) and adopted

by the IUGS and INQUA stratigraphic commissions. Hedberg presented the following hierarchy of lithostratigraphic units:

Unit	Brief outline of definition
Group	consists of two or more formations
Formation	a main unit
Member	a named unit within a formation
Bed	a named unit within a member or formation

Geographical names from the vicinity of the stratigraphic section should in general be used in naming the unit, in addition to the name of the main sediment within the unit (for instance, Chicago Sand Formation, Michigan Silt Member and Green Bay Diamicton Bed). Note that the formal names are written with capital letters and that only descriptive sediment names are used. The Green Bay Diamicton Bed is interpreted as a till bed, but the term "till" involves a genetic interpretation and should not be used in the name.

The term "Formation" should be used for main units that have a considerable lateral extent. This requirement can be hard to fulfill in many cases dealing with Quaternary sediments, where exposures often are in small pits and the different units are relatively thin and often discontinuous. Therefore, some Quaternary geologists have overlooked this requirement, but most of them have used a more informal naming for beds and units observed in limited pits and sections. An example is shown in Fig. 3–54, where the main units are named A, B, C ..., and the sub-units A_1, A_2, A_3 ..., B_1, B_2, B_3, etc.

Biostratigraphy is based on studies of the fossil content of the sediment beds. In biostratigraphy a stratigraphic section is divided into bio-zones, each zone with a characteristic fauna or flora assemblage. According to the stratigraphic code each assemblage zone should be named after the most characteristic or dominating fossils (for instance, the *Betual Zone*, the *Corrylus Zone*, the *Mya Zone*).

However, in northern Europe the old classification with numbered vegetation (pollen) zones is still much used. The zones are used for reference (*see* Fig.

2–68). They correspond with named climate zones, and even the names of the climate zones have been used as names for the vegetation (pollen) zones. (*See* Fig. 2–68 and chronostratigraphy.)

Chronostratigraphy (time stratigraphy) is based on time-fixed boundaries for the units, and the hierarchy of units used is according to Hedberg (1976) as follows:

Chrono-stratigraphy	Geo-chrono-logies	Examples
System	Period	Quaternary
Series	Epoch	Pleistocene, Holocene
Stage	Age	Weichselian, Eemian
Chronozone	Chron	Alleröd, Younger Dryas

In the chronostratigraphic system the sediment sequence deposited during a time interval is considered, and the time interval itself is a geochronology. For instance, the Weichselian Stage is a term which refers to the deposits and events between 115 000 years B.P. and 10 000 years B.P., and the Weichselian Age is the time interval from 115 000 to 10 000 years B.P.

Each chronostratigraphic unit should ideally be defined in a stratotype, but in Quaternary geology this rule has not been strictly followed. For instance, the chronozones for the Holocene and the latest part of the Pleistocene presented in Fig. 2–68 are not defined in stratotypes, but still they are much used. Note that the chronozones in this figure have the same names as the old standard climate/vegetation zones for northern Europe, and they are in fact based more or less on average ages of the climate (vegetation) zones. The boundaries for the vegetation zones are time-transgressive, but the ages of the chronozone boundaries are time-fixed. For instance, the age of the Younger Dryas-Preboreal vegetation-zone boundary is slightly older in Germany-Poland than in Scandinavia, but the age of the Younger Dryas-Preboreal chronozone boundary is exactly 10 000 years B.P. all over the world. Since the same names are used both for chronozones and for climate or vegetation zones, the kind of zone being

used should be indicated (for instance, Younger Dryas chronozone or Younger Dryas vegetation or climate zone). However, some scientists try to avoid using these names for the chronozones, to avoid confusion, but unfortunately there are no good alternative names.

In addition to the three traditional main stratigraphic branches there are several others such as climato-stratigraphy, kineto-stratigraphy, morpho-stratigraphy, isotope-stratigraphy, magneto-stratigraphy, magnetic-susceptibility stratigraphy etc. Some of these branches can be considered as sub-branches of the traditional branches.

Climato-stratigraphy is focussed on the stratigraphic division in climate zones, and various climate indicators are used to define the zones, such as fauna, flora, frost features, soils, glacial deposits etc. Usually the climate zones correspond with the bio-zones.

Kineto-stratigraphy is a special kind of lithostratigraphy based on correlation of tectonic features (folds etc.) which resulted from the activity of glaciers that flowed in various directions. The method has been developed in Denmark, where the flow direction of various Quaternary glacier lobes varied drastically.

Tephro-stratigraphy (tephro-chronology) is the stratigraphy of volcanic ash layers. Some ash layers are well dated and represent characteristic marker beds. *Tephro-chronology* can be very important in volcanic and adjacent regions.

Morpho-stratigraphy is based on correlation of morphological features and the ranging of them in age schemes. For instance, a series of end moraines, a series of raised shorelines etc. can be ranged in morphostratigraphic systems. However, morpho-stratigraphy is not a true stratigraphic discipline.

Lately various qualities of the sediments which were previously unknown or poorly known have been studied and used as important stratigraphic tools. Several of the methods can probably be classified as lithostratigraphic, but they are so special that they will be described separately:

Magneto-stratigraphy is based on the stratigraphic paleomagnetic record. Both the paleomagnetic reversals and

the secular paleomagnetic variations can be used. The paleomagnetic reversals represent dramatic changes of the earth's magnetic field, where the magnetic orientation was completely reversed and the magnetic South Pole became a North Pole. Figure 1–19 shows the stratigraphy of the paleomagnetic reversals. Today this stratigraphy is very much used to date sediment sections in various parts of the world, both marine core sections and loess sections. The secular magnetic variations represent the smaller changes in the earth's magnetic field caused by the gradual shift in the position of the magnetic poles. They have been recorded mainly in stratigraphic sections of lacustrine and shallow marine sediments. However, the method is much used in stratigraphic studies of old rocks also.

Magnetic-susceptibility stratigraphy is based on the recorded changes in magnetic susceptibility in the beds in stratigraphic sections. This method has proved to be important in establishing a stratigraphy in, for instance, loess sections in China, where the magnetic susceptibility in the ice age loess beds is very different from the susceptibility in the interglacial beds (*see* Fig. 1–20A).

Isotope-stratigraphy. The most important isotope stratigraphy is the deep-sea oxygen-isotope stratigraphy (*see* p. 21). The oxygen-isotope zones which are presented in Fig. 1–19 are, in fact, chronozones, and today they are used as reference zones even for terrestrial stratigraphic records. Oxygen isotopes are important for the stratigraphic records of cores from ice sheets and of beds and speleothems also (*see* p. 22).

The carbon isotope ^{13}C has also been used in marine stratigraphy. Apparently the ^{13}C content in ocean water varies with the properties of the water masses, and the ^{13}C content in the fossil marine organisms in a stratigraphic section (core) gives information about the changes.

Carbon content and dissolution. The content of carbon or carbonates in marine deep-sea sediments is usually related to the production of marine organisms. Therefore, the variation in carbon/carbonate content in a core records the variation in marine environment, and it is usually closely related to temperature changes.

The dissolution of carbonate shells in the zone immediately above the carbonate compensation depth, within the lysocline zone, varies with the temperature of the water and the changing depth of the CCD. Therefore, records of the carbonate dissolution may give important stratigraphic information.

stratotype: the same as type section.

stratum: *see* sediment structures.

stress: force per unit area on a body, for instance a rock or a sediment. There are different kinds of stress: compressive, tensile, shear (*see* strain).

strike, strike fault: *see* tectonic structures.

structure of a unit: the sum of all structural features like bedding, folding, faulting, jointing, brecciation, cleavage etc.

subangular: *see* roundness.

Subarctic zone: a climatic zone which lies between the Arctic and the Boreal zones (*see* Fig. 1–28).

subduction zones: *see* plate tectonics.

sublimation: the transition of a solid directly to a vapor, without passing through the liquid state.

sublittoral features: sedimentary features on the sea floor from low-tide level to a depth of about 100 m.

submarine delta: a submarine flat fan, which resembles an ordinary delta, generally deposited at the mouth of a submarine canyon.

submerged area: an area which has dropped relatively to the sea (lake) level and has changed from dry land to be covered by the sea or the lake.

subrounded: *see* roundness.

subsequent stream: a stream that follows a zone of relatively weak rocks, parallel with the bedrock strike.

subsidence: the sinking of a part of the earth's crust.

substage: a time-stratigraphic unit, next lower in rank than a stage.

substratum: a term used both for the rock underlying a sediment cover and for the mobile part of the mantle below the Moho.

superimposed drainage: a drainage which started as consequent drainage on the surface of rock beds, and when the beds were removed by erosion the drainage continued the original course independent of the new structures of the exposed underlying rocks.

surging glacier: a glacier that for a short period moves much faster than normal.

suspended load: *see* load.

Sverdrup: a name used in oceanography for a measure of water-transport in an ocean current. One Sverdrup equals a transport of 1 million m³ per second.

syenite: *see* rock.

synclinal fold: *see* tectonic structures.

syngenetic: a term applied to the formation of minerals, ores or various features which takes place at the same time as the formation of the enclosing rock or sediment.

synorogenic: a term applied to processes etc. which take place during an orogeny.

System: in stratigraphy used as shown on p. 190.

T

taiga: a term used in Russia for the belt of coniferous forest adjacent to the tundra, more or less synonymous with the Boreal forest zone.

talik: a layer of unfrozen ground between frozen units.

talus or scree: the rock material, usually angular clasts, at the foot of a steep slope or cliff, generally with an apron-shaped surface.

taxon (plural: taxa): a named group of organisms which can be a species, a genus, a family etc.

taxonomy: the science of arranging (ranking) living and fossil plants and animals. Phyllum, class, order, family, genus, species and type are the classifications, in descending rank.

tectonic: pertaining to the deformation of the earth's crust resulting in folding, faulting etc.

tectonic structures (*see* Fig. 4–15):

anticline: the opposite of syncline.

diapir: a dome-like sediment body where the sediments have moved up from a source bed and broken through the overlying beds.

dip: the angle between a surface or bed and the horizontal.

dike: a bed-shaped igneous or clastic intrusion which is discordant with the host beds. (*See* clastic dike.)

drag folds: generally small folds formed in incompetent beds when more competent beds on either sides "move".

fault: a break in the rock or sediment along which there has been a displacement, a slip.

flexure: a broad, gentle fold.

graben: a downfaulted belt of the earth's surface, with faults on both sides of the belt.

inclination: (1) the plunge of an axis or dip of a surface; (2) magnetic inclination (*see* p. 178).

inverted: beds that are turned upside down.

isoclinal fold: any fold with parallel limbs.

limb: the limbs of a fold are the straight parts on both sides of the bend.

monocline strata: strata dipping in one direction; they cannot be recognized as a part of a fold.

nappe: a large-scale overfold which has moved as a recumbent fold along a thrust plane for a long distance.

normal fault: see Fig. 4–15.

overthrust: a flat-lying thrust-fault with a relatively large net slip.

overturned fold: see Fig. 4–15.

plunge: the inclination of an axis.

recumbent fold: a fold which has a horizontal or nearly horizontal axial plane.

reverse fault: see Fig. 4–15.

rift valley: a valley formed by a graben.

shear plane: a plane within a rock or sediment along which a slip, caused by shear stress, has taken place.

slip: used in many different connections, often in connection with a fault, where it means the displacement between the two sides of the fault plane.

step fault: see Fig. 4–15.

strike: the strike of a bed, a fault etc. is the direction of the intersection line between a horizontal plane and the bedding plane, fault plane etc.

strike fault: see Fig. 4–15.

synclinal fold: see Fig. 4–15.

thrust or thrust fault: a reverse fault which has a nearly horizontal thrust plane.

telmatic peat: peat accumulated between high and low water levels (*see* terrestrial).

temperate glacier: a glacier where all of the ice is approximately at the pressure melting point at the end of the melting season.

temporary base level: *see* base level.

tephra: a collective term for volcanic material that has been transported through

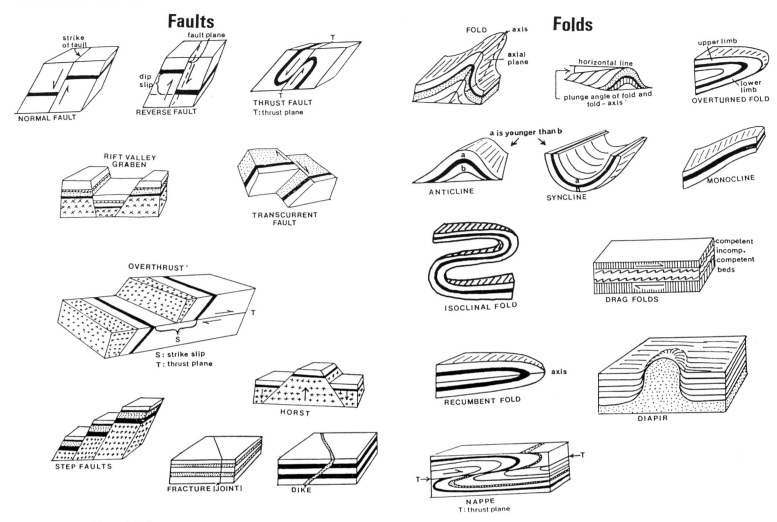

Fig. 4–15. Folds and faults.

the air, such as volcanic ash (not lava).

tephro-chronology: a chronology based on correlating volcanic ash beds (*see* stratigraphy).

terminal moraine: the same as end moraine and front moraine (*see* moraine).

terra rossa: a red soil which is a residuum formed by weathering of calcareous rocks in a warm climate.

terrigenous deposit: deposited on the earth above sea level, as opposed to marine deposits.

Tertiary: *see* Fig. 1–31.

Tethys Sea: the ocean which formed a huge "bay" on the east side of the supercontinent Pangea, 340–225 million years ago.

texture: the term includes the size, shape and arrangement of the particles in a rock or sediment (*see* sediment structures).

thermocarst topography: a bumpy topog-raphy with depressions resulting from the melting of ground ice in permafrost areas (*see* p. 137).

thermograph: an instrument which records temperature changes through time.

thermoluminescence: the emittance of light from some minerals when they are heated, minerals which have stored energy resulting from electron displace-ments in the crystal lattice (*see* p. 148).

thermoluminescence method, TL-method: a method used to date ceramics and even some late Quaternary sediments, based on the TL-process (*see* p. 148).

thermomer: a warm phase, as opposed to a kryomer, a cold phase.

thermophilous species: warmth-demand-ing species.

thrust or thrust fault: *see* tectonic struc-tures.

tidal deltas: delta-like accumulations formed at the mount of tidal inlets/channels by material transported by currents through the inlets/channels.

tide: the periodic rise and fall of the sea level. High tide and low tide are the extremes in a "normal" daily cycle. High spring tide and low spring tide are the extremes which occur when the gravitational forces from the moon and the sun are in phase or in opposite phase, respectively.

ebb tide: the falling tide which occurs during the period after the high tide.

Tiglian: *see* Fig. 2–2.

till: a diamicton carried by, and deposited by, a glacier, generally in contact with the glacier. There are several kinds of till, determined by how the till was deposited.

lodgement till: deposited by lodging, or "plastering on" material from the base of the glacier. The lodgement till usually contains a high percentage of fine-grained clasts, and it is generally very compacted. The elongated clasts usually have a preferred orientation with long axes parallel with the glacier-flow direction, and a preferred dip in upglacier direction.

melt-out till (also called ablation till): deposited when englacial till melts out, in some cases at the base of the glacier, but generally on top of a glacier surface, which can be gradually lowered to the ground by melting. The melt-out till usually contains fewer fine-grained clasts, and it is less compacted than the lodgement till, particularly in the case of surface melt-out. In addition the orientation of the elongated clasts varies considerably, from a preferred orientation parallel with the glacier-flow direction to no preferred orientation. The preferred orientation is most common in subglacial melt-out tills.

flow till: deposited more or less by mudflows of saturated till both in depressions on the glacier and on the ground next to the glacier. Some scientists prefer to use the term mudflow for the last-mentioned kind of flow till.

deformation till: a till of reworked older sediments which are mixed with true glacial debris, and where the original character of the sediment to a certain degree was preserved when the till was deposited. There is a gradational transition from a glacially tectonized sediment, where the original sediment structures are largely preserved, to a deformation till.

waterlaid till: this term has been used for stratified units of diamicton beds or laminae which were deposited in thin water layers below barely floating glaciers. However, it is generally very difficult to distinguish between a waterlaid till and an ice-proximal glaciomarine or glaciolacustrine diamicton.

englacial till: rock debris in transport within the glacier.

The tills compose most of the different kinds of moraines, but not all moraines (*see* moraine).

till fabric: the fabric of tills (*see* fabric analysis).

tillite: a petrified (lithified) till. However, the term has been used in a wider sense which includes petrified glaciofluvial, glaciomarine and periglacial deposits.

time-transgressive: a term used when the age of one lithostratigraphic unit, biostratigraphic zone etc. changes from one area to another (*see* p. 147).

tolerance: the ability of a species to survive variations in its environment.

topset beds: *see* delta.

trace element: an element which occurs in a small, barely detectable quantity in a rock or sediment.

trace fossil: *see* sediment structures.

trade wind: a steady wind at low latitudes, which blows towards the equator, in southwesterly direction in the northern hemisphere.

transgression: the rise of the sea resulting in a gradual submergence of land areas. The cause can be subsidence of the land or eustatic rise of the sea level.

travertine: a hard variety of tufa.

Trene Interstadial: *see* Warthe Stage.

treppen concept: a concept, advocated by Penck, opposing Davies' erosion cycle concept. Davies stressed downwearing and Penck stressed backwasting. According to Penck the uplift and rejuvenation of a land area occurred more or less gradually and the river erosion progressed headwards, forming flat plains which ended in a steep headwall escarpment or steps. New plains and steps were formed one below the other as the uplift continued, and all steps moved gradually headwards.

triangle diagram: a diagram used to present the percentages of three different components in a sample, for instance sand, silt and clay (*see* Fig. 3–58).

Triassic: *see* Fig. 1–31.

triaxial test: a test used to determine when a natural sediment fails under stress (*see* geotechnical parameters).

tributary glacier: *see* trunk glacier.

trim line: a line on the valley (mountain) slope which corresponds with a former glacier surface. Glacial erosion has removed the vegetation below the line, and there is clearly a more established vegetation, or weathering, above the line.

tritium: a radioactive isotope of hydrogen with a short half-life of 12.4 days.

trough: a term used in many different connections; in glacial geology, used for a closed depression with "U-shaped" cross profile. Glacial troughs are the most characteristic forms in glacially sculptured landscapes. They can be

from a few centimeters to several hundred kilometers long and wide, and from a few centimeters to several hundred meters deep. The sides of some troughs are gentle, and some are steep (*see* fjärd and fjord, and p. 114).

truncate: *see* sedimentary structures.

trunk glacier: the main glacier in a valley system where smaller glaciers from tributary valleys join the main glacier.

tschernosem: *see* chernozem soil.

tsuanamy: a large sea wave caused by a submarine earthquake or sudden volcanic eruption.

tufa: a sedimentary rock of calcium carbonate generally deposited from a water solution in a calcareous spring. The term has been used for a similar deposit of silica also, but this use is discouraged.

tuff: a rock of small, volcanic, airborne fragments.

tundra: a treeless plain with sparse or patchy vegetation in a permafrost area.

tunnel valley: a valley formed by erosion of a subglacial river which in some cases flowed upslope towards the ice front. Several of the shallow bays and valleys on the coast of eastern Jutland and northern Germany are supposed to be tunnel valleys (*see* p. 57).

turbidity current: a submarine or sublacustrine density current, consisting of a suspension of sediments, which is heavier than the surrounding water and travels downslope with high speed. Density currents can erode canyons and deposit large, flat submarine fans on the sea floor beyond the mouth of the canyons (*see* Bouma sequence).

type locality: the locality where a type unit or a type stratigraphic border is defined.

type section: a stratigraphic section with a well-defined stratigraphic unit that has become a standard unit, generally located at the place where the unit received its name, i.e. the type locality (*see* stratigraphy).

U

Udden scale: a logarithmic scale for grainsize classification (*see* p. 172).

unconformity: *see* sediment structures.

unconsolidated material: a term generally

used for sediments/soils where no, or very little, consolidation has taken place.

underfit stream: a stream which is much too small to have cut the valley in which it flows.

underlie: to lie beneath. The opposite of overlie.

uniformitarianism: the hypothesis which suggests that the geological processes were the same and produced the same results in the past as in the present ("the present is the key to the past") (*see* catastrophism).

Uranium-series dating: *see* radiometric dating.

Urströmtäl: a German term used for a valley which was formed by a river that flowed along the margin of an ice sheet and today has a course more or less transverse to the general dip of the land surface (*see* Fig. 2–25).

U-shaped valley: *see* glacial-erosion features.

V

Valdaian Glaciation: the last glaciation in the former USSR. The same as the Weichselian Glaciation (*see* Fig. 4–7).

valley glacier: *see* p. 115.

valley train: a long outwash plain in a narrow valley.

Variscan orogeny: a part of the Hercynian orogeny.

varve chronology: a chronology based on counting varves and correlating varve units in different areas. The method was developed in Sweden by De Geer. Annual layers of mainly organic material deposited in lakes have been called organic varves.

varved clay: a laminated clay consisting of dark winter laminae of mainly clay, and lighter silty or sandy summer laminae; each pair is called a varve and represents the deposition during a year. Some varves are several centimeters thick. Varved clay is usually deposited in lakes or in bays which receive water from glacial rivers (*see* sediment structures).

varves: rhythmic couplets where each couplet consists of a summer and a winter layer. The couplets can be inorganic as in varved clay, or they can be mixed organic and inorganic as in sediments in some lakes with less inorganic sediments.

ventifact: a stone which has been polished and shaped by the blasting of wind-blown sand.

vermiculite: *see* clay minerals.

viscosity: internal "friction" in a fluid offering resistance to the flow.

Vistulian Glaciation: the last glaciation in Poland. The same as the Weichselian Glaciation.

V-shaped valley: a valley with a cross profile that resembles a V. This is a typical cross profile for many young river valleys (*see* Fig. 3–14).

W

wady: a generally dry stream channel in a desert area, which carries water only in very short periods with heavy rainfall. The wady frequently lies at or near the foot of mountains from which water drains into the wady during the rainfall.

warping: gentle tilting of a part of the earth's crust.

Warthe Moraine: a young Saalian end moraine in Germany (*see* p. 38).

Warthe Stage: the youngest of the two main Saalian glacial stages, of which the Drenthe Stage is the oldest. Some scientists have considered Warthe and Drenthe to represent full glaciations separated by a Trene Interglacial. However, most scientists seem to oppose this view (*see* Fig. 2–1).

washboard moraines: a series of closely spaced, low moraine ridges which are oriented transversely to the glacier-flow direction, in some cases considered to be annual moraines.

water divide or watershed: a line or narrow zone between two areas in which the water drains to two different streams.

water table: *see* ground water.

wave base: the depth below which there is no effective wave erosion, or transport of material by waves (*see* Fig. 4–14).

weathering: the processes by which rocks and clasts at or near the earth's surface are disintegrated and/or dissolved when exposed to the weather (rain, moisture, wind, frost, heat) and the action of plants and bacteria. It is usual to distinguish between mechanical weathering, which is a mechanical disintegration of rocks and clasts, and chemical weathering.

wedge out: *see* sediment structures.

Weichselian Glacial: *see* Fig. 2–1.

well rounded: *see* roundness.

Wentworth classification: *see* grain size and p. 172.

wind polish: a polish of rock surfaces caused by abrasion from sand grains transported by wind.

wind ripples: *see* sediment structures.

Wisconsin Glacial: *see* Fig. 2–1.

Woodfordian: a term used in Illinois, USA, for the time of maximum glacier extension during the Late Wisconsin.

Würm Glacial: *see* Fig. 2–1.

X – Y – Z

xerophytes: organisms which live under very dry conditions.

XRD: x-ray diffractometer; see DTA.

Yarmouth Interglacial: *see* Fig. 2–1.

Younger Dryas: *see* p. 82.

zone: a term used in several connections, such as:

1. A biostratigraphic zone (*see* stratigraphy).

2. A region on the earth's surface with a defined climate and/or vegetation or animal life, such as the Boreal Zone (*see* p. 30).

3. Informal term for a narrow belt in a bedrock or sediment unit with a characteristic property, for instance a fracture zone.

zone fossil or index fossil: a fossil species which is considered the most characteristic within a fossil assemblage of the zone and usually gives its name to the zone, such as *Betula* Zone.

Recommended Literature

Books
There are so many books dealing with the various aspects of the Quaternary that only a few selected ones have been listed. They are classified as (A): easy to read and (B): best suited for more advanced studies. All listed books are good to excellent.

Dictionaries in geology and physical geography
(A) *Dictionary of Geological Terms* (545 pages), Dolphin Books.
(A) *Dictionary of Physical Geography* (591 pages), Penguin Books.
(A) *The Concise Oxford Dictionary of Earth Sciences* (410 pages), Oxford Univ. Press.
(A) *The Penguin Dictionary of Geology* (500 pages), Penguin Books.
(A) *Longman Illustrated Dictionary of Geology* (192 pages), with colored illustrations.
All of the dictionaries are good and inexpensive. Larger dictionaries are generally available at university libraries.

General Quaternary geology
The classic textbooks by R.F. Flint (*Glacial and Quaternary Geology*, 892 pages, published in 1971), by P. Woldstedt (*Das Eiszeitalter*, Vol. I, II and III, published between 1954 and 1965) and by J.K. Charlesworth (*The Quaternary Era*, Vol. I and II, 591+1700 pages, published in 1957) are now out of print and partly out of date.
The following list contains a few of the many good books which have been published during the last two decades.

(A/B) *The Pleistocene. Geology and Life in the Quaternary Ice Age*, by Tage Nilsson (published in 1983 by Dr. Reidel Publishing Company, London, 651 pages), covers most aspects of the Quaternary, in particularly the biostratigraphical/climatological part. The glacial geological/sedimentological part is somewhat limited.

(A) *Reconstructing Quaternary Environments*, by J.J. Lowe and M.J.C. Walker (published in 1984 by Longman, London, 389 pages), is a good introduction to biostratigraphy, lithostratigraphy, dating methods and climate fluctuations. The regional part is much focussed on Britain, with shorter reviews from other parts of the world.

(A) *Das Eiszeitalter*, by H.D. Kahlke (published in 1981 by Aulis Verlag, Deubner & Co Köln, 191 pages), has many excellent illustrations and pictures in color. It is useful as a brief popular introduction to the Quaternary, but unfortunately the book is available only in the German language.

(A) *Quaternary Geology. A Stratigraphic Framework for Multidisciplinary Work*, by D.O. Bowen (published in 1978 by Pergamon Press, 221 pages), gives a good and short introduction to Quaternary stratigraphy with many examples from British geology, and some from other parts of the world.

(A) *Ice Age Earth. Late Quaternary Geology and Climate*, by A. G. Dawson (published in 1992 by Routledge, London, 293 pages), has introductions to most fields related to the Quaternary glacial and climate fluctuations.

(A) *Quaternary Environments*, by M.A. Williams, D.L. Dunkerley, P. DeDekker, A.P. Kershaw and T. Stokes (published in 1993 by Edward Arnold Ltd, London, 329 pages), covers in particular the Quaternary of the southern hemisphere.

(A) *Ice Ages. Solving the Mystery*, by J. Imbrie and K.P. Imbrie (printed in 1979 by Erslow Publishers, Short Hills, New Jersey, 224 pages), gives a popular introduction to the research history related to the ice age theory and late Cenozoic climate changes.

(A/B) *Quaternary Paleoclimatology. Methods of Paleoclimatic Reconstruction*, by R.S. Bradley (published in 1985 by Allen and Unwin, London, 472 pages), presents good reviews of methods in paleoclimatic reconstruction, dating methods etc.

Glacial and/or periglacial processes
(A) *Glaciers and Landscape*, by D.E. Sugden and B.S. John (published in 1976 by Edward Arnold Ltd, Hill Street, London, 376 pages), deals with glaciers and the results of glacial erosion, transport and deposition.

(B) *The Physics of Glaciers*, by W.S.P. Paterson (published in 1972 by Pergamon Press, Oxford), focusses on the physical-dynamical behavior of glaciers. The mathematical part may be problematic for some readers.

(A) *Geomorphic Responses to Climate Changes*, by W.B. Bull (published in 1991 by Oxford Univ. Press, 321 pages), focusses on different processes, including glacial and periglacial.

(A) *Glacial Geologic Processes*, by D. Drewry (published in 1987 by Edward Arnold Ltd, 276 pages), is a good introduction to glacial-geologic processes.

(A) *Periglacial Geomorphology*, by C. Embleton and A.M. King (published in 1975 by Edward Arnold Ltd, 203 pages), covers periglacial processes and features.

(B) *Geocryology. A Survey of Periglacial Processes and Environment*, by A.L. Washburn (published in 1979 by Edward Arnold Ltd, 406 pages), is an advanced book about frost features and processes, with numerous good illustrations and pictures.

Quaternary mammals
(A) *On the Track of Ice Age Mammals*, by A.J. Sutcliffe (published in 1986 by British Museum of Natural History, London, 224 pages), contains many good pictures and illustrations.

Marine geology
Very much of the information about Quaternary climate and glacier fluctuations was obtained from sediments collected below the ocean floor. There are several good books on marine geology, and many publications are presented in the journal *Marine Geology*. The book listed below deals with glaciomarine sedimentation.

(A) *Glaciomarine Environments. Processes and Sediments*, edited by J.A. Dowdeswell and J.D. Scourse (published in 1990 by The Geological Society, London, 423 pages).

Sedimentology, structural geology and other geological subjects
Numerous textbooks and special books about the various geological subjects that are of importance also in Quaternary geology are usually available at university libraries, and the reader is referred to them.

The Quaternary of restricted regions
Many books cover the Quaternary history of restricted regions or nations. They are usually written in the language native to the region or country.

Journals
The following is a list of some of the best-known international journals dealing with Quaternary aspects. Most of them are available at university libraries.

1) *Quaternary Research*. Issued in USA.
2) *Boreas*. Issued in Scandinavia.
3) *Quaternary Science Reviews*, published by Pergamon Press.
4) *Journal of Quaternary Science*, published by John Wiley and Sons.
5) *Quaternary International*, published by Pergamon Press.
6) *Arctic and Alpine Research*. Issued in USA.
7) *Journal of Glaciology*. Issued in England.
8) *Geografiska Annaler* (Ser. A). Issued in Sweden.
9) *Journal of Biogeography*, published by Blackwell Sci. Publications.
10) *Marine Geology*, published by Elsevier Sci. Publ.
11) *Palaeoecology of Africa*, published by A.A. Balkema, Rotterdam. Many of the papers in this journal deals with Quaternary aspects.
12) *Climatic Change*, published by Kluwer Academic Publications.
13) *Pollen et spores*. Issued in France.

In addition to the listed journals, many other national journals deal with Quaternary topics.

List of Illustrations

References for the Illustrations

Andersen, B., 1981: "Late Weichselian Ice Sheets in Eurasia and Greenland". Pages 3 to 65 in *The Last Great Ice Sheets*, edited by G. Denton and T. Hughes. John Wiley and Sons, New York, 484 pages.

Andersen, B. and Mangerud, J., 1990: "The Last Interglacial-Glacial Cycle in Fennoscandia". *Quaternary International*, Vol. 3/4, pp. 21–29.

Andersen, S. Th., 1965: "Interglacialer og interstadialer i Danmarks kvartær". Medd. dansk geol. forening, Vol. 15, pp. 486–506.

Barnola, J.M., Raynaud, D. Korotkevich, Y.S. and Lorius, C., 1987: "Vostok Ice Core Provides 160 000 Year Record of Atmospheric CO_2". *Nature*, Vol. 329, pp. 408–14.

Behre, K.E., 1989: "Biostratigraphy of the Last Glacial Period in Europe". *Quaternary Science Reviews*, Vol. 8, pp. 25–44.

Benson, L. and Thompson, R.S., 1987: "The Physical Record of Lakes in the Great Basin", pp. 241–60 in *North America and Adjacent Oceans during the Last Deglaciation*, edited by W.F. Ruddiman and H.E. Wright Jr. The Geological Survey of North America, Vol. K-3, 501 pages.

Broecker, W.S., 1987: "The Biggest Chill". *Natural History Magazine*, Vol. 97, pp. 74–83.

Chappell, J. and Shackleton, N.J., 1986: "Oxygen Isotopes and Sea Level". *Nature*, Vol. 324, pp. 137–38.

Coope, G.R., 1977: "Fossil Coleopteran Assemblages as Sensitive Indicators of Climatic Changes during the Devensian (Last) Cold Stage". Phil. Trans. Ray. Soc. London. B. 280, p. 313–40.

Covey, C., 1984: "The Earth's Orbit and the Ice Ages". *Scientific American*. February 1984.

Currey, D.R. and Oviatt, C.G. 1985: "Durations, Average Rates, and Probable Cause of Lake Bonneville Expansion, Still-stands, and Contractions during the Last Deep-lake Cycle, 32 000 to 10 000 Years Ago", pp. 1–9 in *Problems of and Prospects for Predicting Great Salt Lake Levels*, edited by P.A. Kay and H.F. Diaz. Center for Public Affairs and Administrations, Univ. of Utah.

Denton, G. and Hughes, T., 1981: *The Last Great Ice Sheets*. Denton and Hughes (eds.). John Wiley and Sons, New York, 484 pages.

Dreimanis, A. and Karrow, P.F., 1972: "Glacial History of the Great Lakes – St. Lawrence Region, the Classification of the Wisconsinland Stage, and Its Correlatives". International Geol. Congr., 24th Sess. Sec. 12, pp. 5–15.

Dyke, A.S. and Prest, V.K., 1987: "Paleogeography of Northern North America, 18 000–5000 Years Ago". Geological Survey of Canada, Map. 1703A.

Flint, R.F., 1957: *Glacial and Pleistocene Geology*. John Wiley and Sons, 553 pages.

Flint, R.F., 1971: *Glacial and Quaternary Geology*. John Wiley and Sons, 892 pages.

Fuji, N., 1974: "Palynological Investigations on 12-meter and 200-meter Core Samples of Lake Biwa in Central Japan", pp. 227–35 in *Paleolimnology of Lake Biwa and the Japanese Pleistocene*, Shoji Horie (ed.).

Granlund, E. and Lundqvist, G., 1949: "De Kvartära Bildningarna", pp. 212–424 in *Sveriges Geologi*. Editors: N.H. Magnussen, E. Granlund, G. Lundqvist. Svenska Bokförlaget, Stockholm, 424 pages.

Grosswald, M., 1988: Publication in Russian in *Data of Glaciological Studies*, No. 63, pp. 3–25.

Hafsten, U., 1958: "De senkvartaere strandlinjeforskyvningene i Oslotrakten belyst ved pollenanalytiske undersökelser" ("Application of pollen analysis in tracing the Late Quaternary displacement of shorelines in the inner Oslo fjord area"). *Norsk Geografisk Tidsskrift* 16, pp. 74–99.

Hafsten, U., 1983: "Shore-level changes in South Norway during the last 13 000 years, traced by biostratigraphical methods and radiometric datings". *Norsk Geografisk Tidsskrift* 27, pp. 63–79.

Hammer, C.U., Clausen, H.B., Dansgaard, W., Neftel, A., Kristinsdottir, P. and Johnson, E., 1985: "Continuous impurity analysis along the Dye 3 deep core", in *Greenland Ice Core: Geophysics, Geochemistry and the Environment* (eds. C.C. Langway, H. Oeschger and W. Dansgaard), Amer. Geophys. Union, Mon. 33, pp. 90–94.

Harris, S.A., 1969: "The Meaning of Till Fabrics". *Canadian Geographer*, Vol. 13.

Hilgen, T.J., 1991: "Magnetic Reversals Based on Astronimical Calibration of a Cyclic Sedimentary Sequence in Italy". *Earth Planetary Science Letters*, 107, pp. 349–68.

Hollin, J.I. and Schilling, D.H., 1981: "Late Wisconsin-Weichselian Mountain Glaciers and Small Ice Caps", in *The Last Great Ice Sheets*. Edited by G.H. Denton and T.J. Hughes. John Wiley and Sons, 484 pages.

Holtedahl, O., 1940: *The submarine relief off the Norwegian coast.* Det Norske Videnskaps-Akademi i Oslo, 43 pages.

Hopkins, D.M., 1967: *The Bering Land Bridge,* Stanford Univ. Press.

Hunt, C.B., 1972: Fig. 7.8, p. 139, in *Geology of Soils.* W.H. Freeman and Company, San Francisco, 344 pages.

Imbrie, J. et al., 1984: "The Orbital Theory of Pleistocene Climate; Support from a Revised Chronology of the Marine δ¹⁸O Record", in Berger, A., Imbrie, J., Hays, J., Kukla, G. and Saltzman, B., eds. *Milankovitch and Climate*, Part 1, Dordrecht, Reidel, pp. 269–305.

Jouzel, J., Lorius, C., Petit, J.R., Genthon, C., Barkov, N.I. Kotlyakov, V.M. and Petrov, V.M., 1987: "Vostok Ice Core: A Continuous Isotope Temperature Record over the Last Climatic Cycle (160 000 years)". *Nature,* Vol. 329, pp. 403–8.

Kakkuri, J., 1993: "The Stress Phenomenon in the Fennoscandian Shield", pp. 71–86 in *Geodesy and Geophysics.* J. Kakuri (ed). Publ. Finnish Geod. Inst. 115.

Kellogg, T.B., 1976: "Late Quaternary Climate Changes: Evidence from Deep-Sea Cores of Norwegian Greenland Seas". Geol. Soc. of Am. Memoir, Vol. 145, pp. 77–110.

Kristoffersen, Y., Elverhøi, A. and Vinje, T., 1978: Barentshavprosjektet, marin geofysikk, geologi og havis. Unpublished report. Norsk Polarinst. 81 pages.

Kuhle, M., 1988: "The Pleistocene Glaciation of Tibet and the Onset of Ice Ages – An Autocycle Hypothesis". *Geo. Journal* 17.4, pp. 581–95.

Kukla, G., 1987: Loess Stratigraphy in Central China. *Quaternary Science Reviews,* Vol. 6.

Kurten, B., 1968: *Pleistocene Mammals of Europe.* Aldine Publ. Co., Chicago.

MacClintok, P. and Richards, H.G., 1936: "Correlation of the Late Pleistocene Marine and Glacial Deposits of New Jersey and New York". *Bull Amer.,* Vol. 47.

Marthinussen, M., 1960: Maps presented in Holtedahl, 1960: Geology of Norway. Norwegian Geol. Survey Skr., No. 208.

Martinson, D.G., Pisias, N.G., Hays, J.D., Imbrie, J., Moore, T.C. and Shackleton, N.J., 1987: "Age Dating and the Orbital Theory of Ice Ages: Development of a High-Resolution 0 to 300 000-Year Chronostratigraphy". *Quaternary Research* 27, pp. 1–30.

Menke, B. and Tynni, R., 1984: "Das Eeminterglazial und das Weichselfrühglazial von Rederstall/Ditmarschen und ihre Bedeutung für die mitteleuropäische Jungpleistozän-Gliederungen". *Geologisches Jahrbuch Reihe A,* Vol. 76, pp. 3–120.

Neef, E., 1970: *Das Gesicht der Erde.* Harri Deutsch Verlag Frankfurt and Zürich.

Oeshger, H. and Langway, C., 1989: *The Environmental Record in Glaciers and Ice Sheets.* A. Wiley Interscience Publ., 1989, New York.

Olsson, I.U., 1974: "Some Problems in Connection with the Evaluation of C¹⁴ Dates". Geol. Föreningen i Stockh. Förh., Vol. 96, pp. 311–20.

Paterson, W.S.B. and Hammer, C.U., 1987: "Ice Core and other Glaciological Data", pp. 91–109 in *North America and Adjacent Oceans: During the Last Deglaciation.* W.F. Ruddiman and H.E. Wright Jr. (eds.), *The Geology of North America,* Vol. K-3, Geol. Soc. Am.

Penck, A. and Brückner, E., 1901–1909: *De Alpen im Eiszetalter,* 1–3. Leipzig.

Shackleton, N.J. and Opdyke, N.D., 1976: "Oxygen Isotope and Palaeomagnetic Stratigraphy of Pacific Core V28-239, Late Pliocene to Late Holocene", Geol. Soc. Am. Mem., Vol. 145, pp. 449–64.

Shackleton, N.J., Backman, J., Zimmermann, H., Kent, D.V., Hall, M.A., Roberts, D.G., Schnitker, D., Baldauf, J.G., Desprairies, A., Homrighausen, R., Huddlestun, P., Keene, J.B., Kaltenback, A.J., Krumsiek, K.A.O., Morton, A.C., Murray, J.W. and Westberg-Smith, J., 1984: "Oxygen-Isotope Calibration of the Onset of Ice-Rafting and History of Glaciation in the North Atlantic Region". *Nature*, Vol. 307, pp. 620–23.

Shackleton, N.J., Berger, A. and Pettier, W.R., 1990: "An Alternative Astronimical Calibration of the Lower Pleistocene Timescale Based on ODP Site 677". Transaction of the Royal Society of Edinburgh: *Earth Sciences* 81, pp. 251–61.

Shackleton, N.J., Hall, M.A. and Pate, D., 1993: "Pliocene Stable Isotope Stratigraphy of ODP Site 846", in *Proc. of ODP Scientific Results*, Vol. 138. Eds: N. Pisias, L. Mayer, T. Janacek et al. (in press).

Sigenthaler, U., Eicher, U., Oeschger, H. and Dansgaard, W., 1984: "Lake Sediments as Continental δ^{18}O Records from the Glacial/Post-Glacial Transition". *Annals of Glaciology*, Vol. 5, pp. 149–52.

Street, F.A. and Grove, A.T., 1979: "Global Maps of Lake-Level Fluctuations since 30 000 BP". *Quaternary Research*, Vol. 12, pp. 83–118.

Ström, K., 1936: *Land-Locked Waters*. Norske Videnskaps-Akad. Skr. I, No. 7.

Stuiver, M. and Reimer, P.J., 1986: "A Computer Program for Radiocarbon Age Calibration". *Radiocarbon*, Vol. 28, No. 2B, pp. 1022–30.

Svendsen, J. and Mangerud, J., 1987: "Late Weichselian and Holocene Sea-Level History for a Cross Section of Western Norway". *Journ. Quat. Sci.*, Vol. 2, pp. 113–32.

Teller, J.T., 1987: "Proglacial Lakes and the Southern Margin of the Laurentide", pp. 39–69 in *North America and Adjacent Oceans during the Last Deglaciation*. Eds.: W.F. Ruddiman and H.E. Wright Jr. The Geological Society of America Inc., 501 pages.

Waga, J.M., 1987: "Fossil Fissure Structures in the Loesses Adjacent to Kazimierza Wielka". *Geographia. studia et dissertationes.* T 10 Katowice. Prace Naukowe Uniw. Slaskiego, No. 813, pp. 7–17.

Williams, I.M., Chen, Y.H., Van der Veen, C.J. and Huges, T.J., 1989: "Force Budget III: Application to the Three-Dimensional Flow of Byrd Glacier". *Antarctica. Journ. Glaci.*, Vol. 35, No. 119, pp. 68–80.

Winograd, I.J., Coplen, T.B., Landwehr, J.M., Riggs, A.C., Ludwig, K.R., Szabo, B.J., Kolesar, P.T. and Revesz, K.M., 1992: "Continuous 500 000-Year Climate Record from Vein Calcite in Devils Hole, Nevada". *Science*, Vol. 258, pp. 255–60.

Woillard, G.M., 1978: "Grande Pile Peat Bog: A Continuous Pollen Record for the Last 140 000 Years". *Quaternary Research*, Vol. 9, pp. 1–21.

Woillard, G.M., 1982: "Carbon-14 Dates at Grande Pile: Correlation of Land and Sea Chronologies". *Science*, Vol. 215, pp. 159–61.

Woldstedt, P., 1954: *Norddeutschland und angrenzende gebiete in Eiszeitalter*. K.F. Koehler Verlag, Stuttgart.

Woldstedt, P., 1954: "Die Klimakurve des Tertiärs und quartärs in Mitteleuropa". *Eiszeitalter und Gegenwart*, Bd. 4/5, pp. 5–9.

Woldstedt, P. and Duphorn, K., 1974: *Norddeutschland und Angrenzende Gebiete im Eiszeitalter*. K.F. Koehler Verlag, Stuttgart, 499 pages.

Zagwijn, W.H., 1975: "Variations in Climate as Shown by Pollen Analysis, Especially in the Lower Pleistocene of Europe", pp. 137–52 in A.E. Wright, J. Moseley and G. Newall (eds.) *Ice Ages: Ancient and Modern*. Geol. Jour. Spec. Iss., No. 6.

Zagwijn, W.H., 1985: "An Outline of the Quaternary Stratigraphy of the Netherlands". *Geologie en Mijnbouw*, Vol. 64, pp. 17–24.

Photographs and drawings at which no reference is cited are in general taken/drawn/colored by the senior author.

Extended Table of Contents

Chapter 3
Processes and
Scientific Methods ....

Index

Scientific terms are listed in the Extended Glossary. In addition, most of the main terms occur in the Extended Table of Contents and in the List of Illustrations. Therefore, this index consists mainly of names of places and scientists. The following abbreviations are used in geographical names:

Ch.   – Channel
Gl.   – Glacier
I.   – Island
M.   – Moraine
Mts   – Mountains

In this book, we have tried to avoid using the Scandinavian letters å, æ and ø. The Danish, Norwegian and Swedish letter å is written aa and alphabetized correspondingly. The Danish and Norwegian letter æ is written ae and alphabetized correspondingly. The Finnish, German and Swedish letter ä is alphabetized as a. The Danish and Norwegian letter ø is written as the Finnish, German and Swedish letter ö, which is alphabetized as o. The German letter ü is alphabetized as u.